MIND
AND
NATURE

BOOKS BY GREGORY BATESON

BALINESE CHARACTER: A PHOTOGRAPHIC ANALYSIS. *Special Publications of the New York Academy of Sciences, vol. 2. New York: New York Academy of Sciences, 1942. With Margaret Mead.*

COMMUNICATION: THE SOCIAL MATRIX OF PSYCHIATRY. *New York: W. W. Norton, 1951. With Jurgen Ruesch.*

NAVEN: A SURVEY OF THE PROBLEMS SUGGESTED BY A COMPOSITE PICTURE OF THE CULTURE OF A NEW GUINEA TRIBE DRAWN FROM THREE POINTS OF VIEW. *Cambridge: Cambridge Univ. Press, 1936. 2d ed., with "Epilogue 1958." Stanford: Stanford Univ. Press, 1965.*

PERCEVAL'S NARRATIVE: A PATIENT'S ACCOUNT OF HIS PSYCHOSIS, 1830–1832, *by John Perceval. Edited with an Introduction by Gregory Bateson. Stanford: Stanford Univ. Press, 1961.*

STEPS TO AN ECOLOGY OF MIND: COLLECTED ESSAYS IN ANTHROPOLOGY, PSYCHIATRY, EVOLUTION, AND EPISTEMOLOGY. *New York: Chandler, 1972; Paperback edition, Ballantine Books, 1972.*

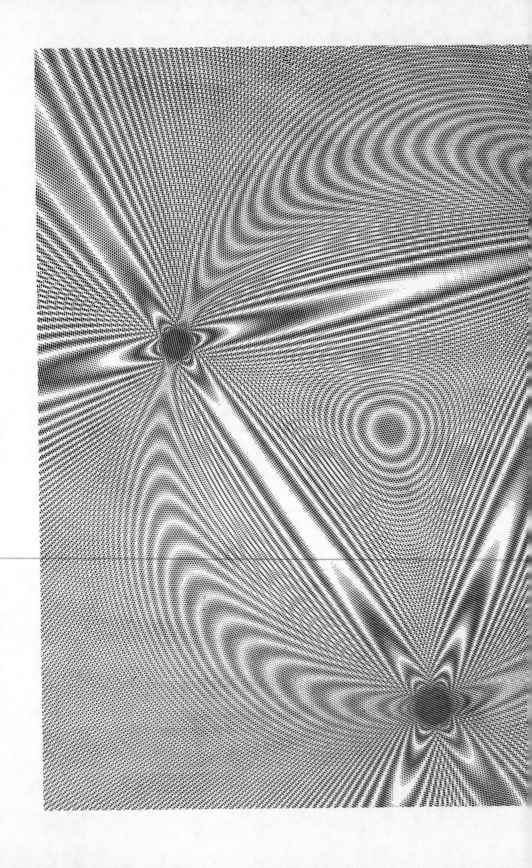

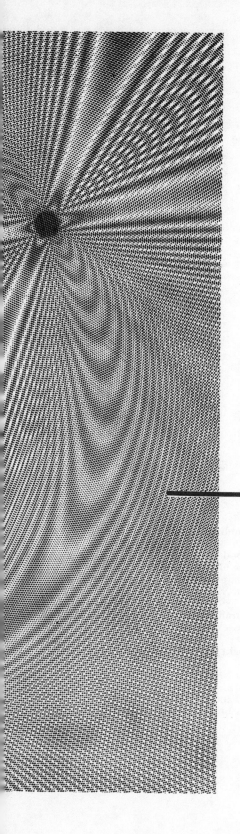

MIND
AND
NATURE

A Necessary Unity

Gregory Bateson

WILDWOOD HOUSE LONDON
BOOKWISE AUSTRALIA

First published in Great Britain 1979

Wildwood House Limited
1 Prince of Wales Passage
117 Hampstead Road
London NW1 3EE

In Australia by Bookwise Australia Pty Ltd
104 Sussex Street, Sydney, 2000

ISBN 0 7045 3014 7

Printed and bound in Great Britain by
Biddles Ltd, Guildford

For:

Nora
Vanni
Gregory
Emily Elizabeth

CONTENTS

ACKNOWLEDGMENTS

The work and thought leading to this book have spread over many years, and my debts go back to include all that were acknowledged in the preface to my previous book, *Steps to an Ecology of Mind*. But I have tried to write to be understandable to those who have not read *Steps* and therefore shall acknowledge here only debts contracted since *Steps* was published.

Even so, recent favors have been many. In something like chronological order, I have to thank first the fellowship of the University of California at Santa Cruz and especially my friends in Kresge College: Mary Diaz, Robert Edgar, Carter Wilson, Carol Proudfoot, and the secretariat.

And then I have to thank the Lindisfarne Association, whose scholar in residence I was for six months of the writing of this book. Bill Irwin Thompson, Michael Katz, Nina Hagen, and Chris and Diane Bamford were hosts who combined generosity with brains. Without them, there would have been no book.

Similarly, in the last stages of writing the book and following severe medical adventures, Esalen Institute took me in as guest, permitting me to combine writing with convalescence. I have to thank Janet Lederman, Julian Silverman, Michael Murphy, Richard Price, and many others. Both at Esalen and at Lindisfarne, my debt is really to the total community.

Early in 1978, I underwent major surgery and was warned that time might be short. In this emergency, Stewart Brand and the Point Foundation came to my aid. Stewart made it possible for my daughter Mary Catherine to come from Tehran and spend a month with me in California working on the manuscript. Her employer in Iran, the Reza Shah Kabir University, generously gave her a professional leave. The first five chapters of the book owe much to her clarifying criticism and sheer hard work. I also thank Stewart for publishing parts of the manuscript in *Co-evolution Quarterly* and for permitting republication here.

Two students of mine have been active and constructive critics, Rodney Donaldson and David Lipset; many others, by listening, have helped me to hear when I was talking nonsense.

My editor, Bill Whitehead, and agent, John Brockman, have patiently nagged me into getting the book written.

My secretary, Judith Van Slooten, did much of the drudgery and helped compile the index, and many others at Lindisfarne and Esalen and along the way have helped.

Finally, my wife, Lois, stood by, criticized and appreciated, and bore patiently with my varying excitements and depressions as the ideas came and went.

GREGORY BATESON was born in 1904, the son of William Bateson, a leading British biologist and a pioneering geneticist. Resisting family pressures to follow in his father's footsteps, he completed his degree in anthropology instead of the natural sciences, and left England to do field work in New Guinea. It was on his second trip there, in 1936, that he met his fellow anthropologist Margaret Mead, whom he later married; their only child, Mary Catherine Bateson, is also an anthropologist. Bateson and Mead were divorced in 1950, but they continued to collaborate professionally and maintained their friendship until Mead's death in 1978.

In the years to follow, Bateson became a visiting professor of anthropology at Harvard (1947); was appointed research associate at the Langley Porter Neuropsychiatric Institute in San Francisco; worked as Ethnologist at the Palo Alto Veterans Administration Hospital (where he developed the double-bind theory of schizophrenia and formulated a new theory of learning). He worked with dolphins at the Oceanographic Institute in Hawaii and taught at the University of Hawaii. In 1972 he joined the faculty of the University of California at Santa Cruz.

The author of *Naven* and *Steps to an Ecology of Mind,* and co-author of *Balinese Character,* Gregory Bateson has markedly influenced an entire generation of social scientists, including the British psychiatrist R. D. Laing—and he is considered one of the "fathers" of the family therapy movement. Appointed by Governor Jerry Brown as a member of the Board of Regents of the University of California in 1976, he now lives in Ben Lomond, California, with his wife, Lois, and daughter, Nora.

MIND
AND
NATURE

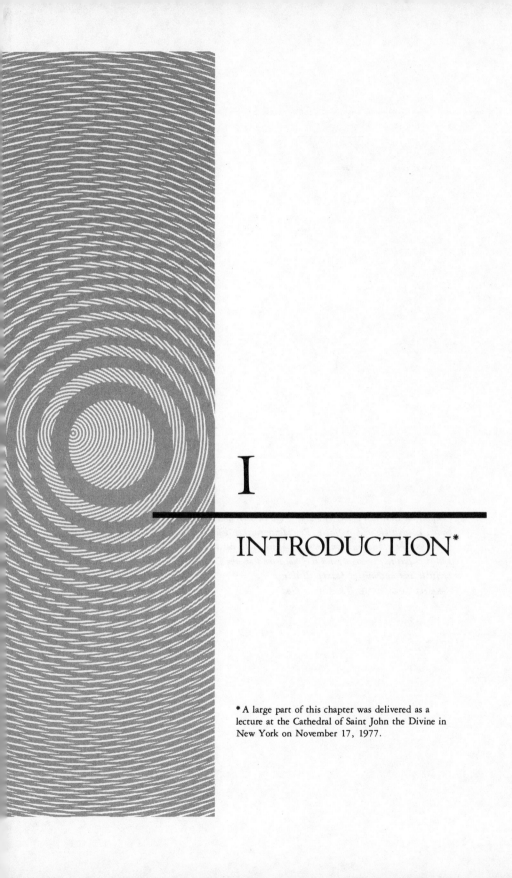

I

INTRODUCTION*

* A large part of this chapter was delivered as a
lecture at the Cathedral of Saint John the Divine in
New York on November 17, 1977.

Plotinus the Platonist proves by means of the blossoms and leaves that from the Supreme God, whose beauty is invisible and ineffable, Providence reaches down to the things of earth here below. He points out that these frail and mortal objects could not be endowed with a beauty so immaculate and so exquisitely wrought, did they not issue from the Divinity which endlessly pervades with its invisible and unchanging beauty all things.
—SAINT AUGUSTINE, *The City of God*

In June 1977, I thought I had the beginnings of two books. One I called *The Evolutionary Idea* and the other *Every Schoolboy Knows.** The first was to be an attempt to reexamine the theories of biological evolution in the light of cybernetics and information theory. But as I began to write that book, I found it difficult to write with a real audience in mind who, I could hope, would understand the formal and therefore simple presuppositions of what I was saying. It became monstrously evident that schooling in this country and in England and, I suppose, in the entire Occident was so careful to avoid all crucial issues that I would have to write a second book to explain what seemed to me

---

* A favorite phrase of Lord Macaulay's. He is credited with, "Every schoolboy knows who imprisoned Montezuma, and who strangled Atahualpa."

elementary ideas relevant to evolution and to almost any other biological or social thinking—to daily life and to the eating of breakfast. Official education was telling people almost nothing of the nature of all those things on the seashores and in the redwood forests, in the deserts and the plains. Even grown-up persons with children of their own cannot give a reasonable account of concepts such as entropy, sacrament, syntax, number, quantity, pattern, linear relation, name, class, relevance, energy, redundancy, force, probability, parts, whole, information, tautology, homology, mass (either Newtonian or Christian), explanation, description, rule of dimensions, logical type, metaphor, topology, and so on. What are butterflies? What are starfish? What are beauty and ugliness?

It seemed to me that the writing out of some of these very elementary ideas could be entitled, with a little irony, *"Every Schoolboy Knows."*

But as I sat in Lindisfarne working on these two manuscripts, sometimes adding a piece to one and sometimes a piece to the other, the two gradually came together, and the product of that coming together was what I think is called a *Platonic* view.* It seemed to me that in *"Schoolboy,"* I was laying down very elementary ideas about *epistemology* (see Glossary), that is, about *how we can know anything.* In the pronoun *we,* I of course included the starfish and the redwood forest, the segmenting egg, and the Senate of the United States.

And in the *anything* which these creatures variously know, I included "how to grow into five-way symmetry," "how to survive a forest fire," "how to grow and still stay the same shape," "how to learn," "how to write a constitution," "how to invent and drive a car," "how to count to seven," and so on. Marvelous creatures with almost miraculous knowledges and skills.

Above all, I included "how to evolve," because it seemed to me that both evolution and learning must fit the same formal regularities or so-called laws. I was, you see, starting to use the ideas of *"Schoolboy"* to

---

* Plato's most famous discovery concerned the "reality" of ideas. We commonly think that a dinner plate is "real" but that its circularity is "only an idea." But Plato noted, first, that the plate is not truly circular and, second, that the world can be perceived to contain a very large number of objects which simulate, approximate, or strive after "circularity." He therefore asserted that "circularity" is *ideal* (the adjective derived from *idea*) and that such ideal components of the universe are the real explanatory basis for its forms and structure. For him, as for William Blake and many others, that "Corporeal Universe" which our newspapers consider "real" was some sort of spin-off from the truly real, namely the forms and ideas. In the beginning was the idea.

reflect, not upon our own knowing, but upon that *wider knowing* which is the glue holding together the starfishes and sea anemones and redwood forests and human committees.

My two manuscripts were becoming a single book because there is a single knowing which characterizes evolution as well as *aggregates* of humans, even though committees and nations may seem stupid to two-legged geniuses like you and me.

I was transcending that line which is sometimes supposed to enclose the human being. In other words, as I was writing, mind became, for me, a reflection of large parts and many parts of the natural world outside the thinker.

On the whole, it was not the crudest, the simplest, the most animalistic and primitive aspects of the human species that were reflected in the natural phenomena. It was, rather, the more complex, the aesthetic, the intricate, and the elegant aspects of people that reflected nature. It was not my greed, my purposiveness, my so-called "animal," so-called "instincts," and so forth that I was recognizing on the other side of that mirror, over there in "nature." Rather, I was seeing there the roots of human symmetry, beauty and ugliness, aesthetics, the human being's very aliveness and little bit of wisdom. His wisdom, his bodily grace, and even his habit of making beautiful objects are just as "animal" as his cruelty. After all, the very word "animal" means "endowed with mind or spirit (*animus*)."

Against this background, those theories of man that start from the most animalistic and maladapted psychology turn out to be improbable first premises from which to approach the psalmist's question: "Lord, What is man?"

I never could accept the first step of the Genesis story: "In the beginning the earth was without form and void." That primary *tabula rasa* would have set a formidable problem in thermodynamics for the next billion years. Perhaps the earth never was any more a *tabula rasa* than is, a human zygote—a fertilized egg.

It began to seem that the old-fashioned and still-established ideas about epistemology, especially human epistemology, were a reflection of an obsolete physics and contrasted in a curious way with the little we seem to know about living things. It was as if members of the species, man, were supposed to be totally unique and totally materialistic against the background of a living universe which was generalized (rather than unique) and spiritual (rather than materialistic).

There seems to be something like a Gresham's law of cultural evolution according to which the oversimplified ideas will always displace the sophisticated and the vulgar and hateful will always displace the beautiful. And yet the beautiful persists.

It began to seem as if organized matter—and I know nothing about unorganized matter, if there be any—in even such a simple set of relations as exists in a steam engine with a governor was wise and sophisticated compared with the picture of human spirit that orthodox materialism and a large part of orthodox religion currently drew.

The germ of these ideas had been in my mind since I was a boy. But let me start from two contexts in which these thoughts began to insist on utterance: In the 1950s, I had two teaching tasks. I was teaching psychiatric residents at a Veterans Administration mental hospital in Palo Alto and young beatniks in the California School of Fine Arts in San Francisco. I want to tell you how those two courses commenced, how I approached those two contrasting audiences. If you put these two first lectures side by side, you will see what I am trying to say.

To the psychiatrists, I presented a challenge in the shape of a small exam paper, telling them that by the end of the course they should understand the questions in it. Question 1 asked for brief definitions of (a) "sacrament" and (b) "entropy."

The young psychiatrists in the 1950s were, in general, unable to answer *either* question. Today, a few more could begin to talk about entropy (see Glossary). And I suppose there are still some Christians who could say what a sacrament is?

I was offering my class the core notions of 2,500 years of thought about religion and science. I felt that if they were going to be doctors (medical doctors) of the human soul, they should at least have a foot on each side of the ancient arguments. They should be familiar with the central ideas of both religion and science.

For the art students, I was more direct. It was a small group of about ten to fifteen students, and I knew that I would be walking into an atmosphere of skepticism bordering on hostility. When I entered it was clear that I was expected to be an incarnation of the devil, who would argue for the common sense of atomic warfare and pesticides. In those days (and even today?), science was believed to be "value-free" and not guided by "emotions."

I was prepared for that. I had two paper bags, and the first of these I opened, producing a freshly cooked crab, which I placed on the table. I then challenged the class somewhat as follows: "I want you to produce arguments which will convince me that this object is the remains of a living thing. You may imagine, if you will, that you are Martians and that on Mars you are familiar with living things, being indeed yourselves alive. But, of course, you have never seen crabs or lobsters. A number of objects like this, many of them fragmentary, have arrived, perhaps by meteor. You are to inspect them and arrive at the conclusion that they are the remains of living things. How would you arrive at that conclusion?"

Of course, the question set for the psychiatrists was the *same question* as that which I set for the artists: Is there a biological species of entropy?

Both questions concerned the underlying notion of a dividing line between the world of the living (where *distinctions* are drawn and *difference* can be a cause) and the world of nonliving billiard balls and galaxies (where forces and impacts are the "causes" of events). These are the two worlds that Jung (following the Gnostics) calls *creatura* (the living) and *pleroma* (the nonliving).* I was asking: What is the difference between the physical world of pleroma, where forces and impacts provide sufficient basis of explanation, and the *creatura,* where nothing can be understood until *differences* and *distinctions* are invoked?

In my life, I have put the descriptions of sticks and stones and billiard balls and galaxies in one box, the pleroma, and have left them alone. In the other box, I put living things: crabs, people, problems of beauty, and problems of difference. The contents of the second box are the subject of this book.

I was griping recently about the shortcomings of occidental education. It was in a letter to my fellow regents of the University of California, and the following phrase crept into my letter:

* C. G. Jung, *Septem Sermones ad Mortuos* (London: Stuart & Watkins, 1967).

"Break the pattern which connects the items of learning and you necessarily destroy all quality."

I offer you the phrase *the pattern which connects* as a synonym, another possible title for this book.

The pattern which connects. Why do schools teach almost nothing of the pattern which connects? Is it that teachers know that they carry the kiss of death which will turn to tastelessness whatever they touch and therefore they are wisely unwilling to touch or teach anything of real-life importance? Or is it that they carry the kiss of death *because* they dare not teach anything of real-life importance? What's wrong with them?

What pattern connects the crab to the lobster and the orchid to the primrose and all the four of them to me? And me to you? And all the six of us to the amoeba in one direction and to the back-ward schizophrenic in another?

I want to tell you why I have been a biologist all my life, what it is that I have been trying to study. What thoughts can I share regarding the total biological world in which we live and have our being? How is it put together?

What now must be said is difficult, appears to be quite *empty,* and is of very great and deep importance to you and to me. At this historic juncture, I believe it to be important to the survival of the whole biosphere, which you know is threatened.

What is the pattern which connects all the living creatures?

Let me go back to my crab and my class of beatniks. I was very lucky to be teaching people who were not scientists and the bias of whose minds was even antiscientific. All untrained as they were, their bias was aesthetic. I would define that word, for the moment, by saying that they were *not* like Peter Bly, the character of whom Wordsworth sang

A primrose by the river's brim
A yellow primrose was to him;
And it was nothing more.

Rather, they would meet the primrose with *recognition* and *empathy.* By *aesthetic,* I mean responsive to *the pattern which connects.* So you see, I was lucky. Perhaps by coincidence, I faced them with what was (though I knew it not) an aesthetic question: *How are you related to this creature? What pattern connects you to it?*

By putting them on an imaginary planet, "Mars," I stripped them of all thought of lobsters, amoebas, cabbages, and so on and forced the diagnosis of life back into identification with living self: *"You* carry the bench marks, the criteria, with which you could look at the crab to find that it, too, carries the same marks." My question was much more sophisticated than I knew.

So they looked at the crab. And first of all, they came up with the observation that it is *symmetrical;* that is, the right side resembles the left.

"Very good. You mean it's *composed,* like a painting?" (No response.)

Then they observed that one claw was bigger than the other. So it was *not* symmetrical.

I suggested that if a number of these objects had come by meteor, they would find that in almost all specimens it was the same side (right or left) that carried the bigger claw. (No response. "What's Bateson getting at?")

Going back to symmetry, somebody said that *"yes, one claw is bigger than the other, but both claws are made of the same parts."*

Ah! What a beautiful and noble statement that is, how the speaker politely flung into the trash can the idea that *size* could be of primary or profound importance and went after the *pattern which connects.* He discarded an asymmetry in size in favor of a deeper symmetry in formal relations.

Yes, indeed, the two claws are characterized (ugly word) by embodying *similar relations between parts.* Never quantities, always shapes, forms, and relations. This was, indeed, something that characterized the crab as a member of *creatura,* a living thing.

Later, it appeared that not only are the two claws built on the same "ground plan," (i.e., upon corresponding sets of relations between corresponding parts) but that these relations between corresponding parts extend down the series of the walking legs. We could recognize in every leg pieces that corresponded to the pieces in the claw.

And in your own body, of course, the same sort of thing is true.

Humerus in the upper arm corresponds to femur in the thigh, and radius-ulna corresponds to tibia-fibula; the carpals in the wrist correspond to tarsals in the foot; fingers correspond to toes.

The anatomy of the crab is repetitive and rhythmical. It is, like music, repetitive with modulation. Indeed, the direction from head toward tail corresponds to a sequence in time: In embryology, the head is older than the tail. A flow of information is possible, from front to rear.

Professional biologists talk about phylogenetic *homology* (see Glossary) for that *class* of facts of which one example is the formal resemblance between my limb bones and those of a horse. Another example is the formal resemblance between the appendages of a crab and those of a lobster.

That is one class of facts. Another (somehow similar?) class of facts is what they call *serial homology*. One example is the rhythmic repetition with change from appendage to appendage down the length of the beast (crab or man); another (perhaps not quite comparable because of the difference in relation to time) would be the bilateral symmetry of the man or crab.*

Let me start again. The parts of a crab are connected by various patterns of bilateral symmetry, of serial homology, and so on. Let us call these patterns *within* the individual growing crab *first-order connections*. But now we look at crab and lobster and we again find connection by pattern. Call it *second-order connection,* or phylogenetic homology.

Now we look at man or horse and find that, here again, we can see symmetries and serial homologies. When we look at the two together, we find the same cross-species sharing of pattern with a difference (phylogenetic homology). And, of course, we also find the same discarding of magnitudes in favor of shapes, patterns, and relations. In

---

* In the serial case it is easy to imagine that each anterior segment may give information to the next segment which is developing immediately behind it. Such information might determine orientation, size, and even shape of the new segment. After all, the anterior is also antecedent in time and could be the quasi-logical antecedent or model for its successor. The relation between anterior and posterior would then be asymmetrical and complementary. It is conceivable and even expectable that the symmetrical relation between right and left is doubly asymmetrical, i.e., that each has some complementary control over the development of the other. The pair would then constitute a circuit of *reciprocal* control. It is surprising that we have almost no knowledge of the vast system of communication which must surely exist to control growth and differentiation.

other words, as this distribution of formal resemblances is spelled out, it turns out that gross anatomy exhibits three levels or logical types of descriptive propositions:

1. The parts of any member of *Creatura* are to be compared with other parts of the same individual to give first-order connections.

2. Crabs are to be compared with lobsters or men with horses to find similar relations between parts (i.e., to give second-order connections).

3. The *comparison* between crabs and lobsters is to be compared with the comparison between man and horse to provide third-order connections.

We have constructed a ladder of how to think about—about what? Oh, yes, the pattern which connects.

My central thesis can now be approached in words: The *pattern which connects is a metapattern.* It is a pattern of patterns. It is that metapattern which defines the vast generalization that, indeed, *it is patterns which connect.*

I warned some pages back that we would encounter emptiness, and indeed it is so. Mind is empty; it is no-thing. It exists only in its ideas, and these again are no-things. Only the ideas are immanent, embodied in their examples. And the examples are, again, no-things. The claw, *as an example,* is not the *Ding an sich;* it is precisely *not* the *"thing in itself."* Rather, it is what mind makes of it, namely, an *example* of something or other.

Let me go back to the classroom of young artists.

You will recall that I had *two* paper bags. In one of them was the crab. In the other I had a beautiful large conch shell. By what token, I asked them, could they know that the spiral shell had been part of a living thing?

When she was about seven, somebody gave my daughter Cathy a cat's-eye mounted as a ring. She was wearing it, and I asked her what it was. She said it was a cat's-eye.

I said, "But what *is* it?"

"Well, I know it's not the eye of a cat. I guess it's some sort of stone."

I said, "Take it off and look at the back of it."

She did that and exclaimed, "Oh, it's got a spiral on it! It must have belonged to something alive."

Actually, these greenish disks are the opercula (lids) of a species of tropical marine snail. Soldiers brought lots of them back from the Pacific at the end of World War II.

Cathy was right in her major premise that all spirals in this world except whirlpools, galaxies, and spiral winds are, indeed, made by living things. There is an extensive literature on this subject, which some readers may be interested in looking up (the key words are *Fibonacci series* and *golden section*).

What comes out of all this is that a spiral is a figure that *retains its shape (i.e., its proportions) as it grows* in one dimension by addition at the open end. You see, there are no truly static spirals.

But the class had difficulty. They looked for all the beautiful formal characteristics that they had joyfully found in the crab. They had the idea that formal symmetry, repetition of parts, modulated repetition, and so on were what teacher wanted. But the spiral was *not* bilaterally symmetrical; it was not segmented.

They had to discover (a) that all symmetry and segmentation were somehow a result, a payoff from, the fact of growth; and (b) that growth makes its formal demands; and (c) that one of these is satisfied (in a mathematical, an ideal, sense) by spiral form.

So the conch shell carries the snail's *prochronism*—its record of how, *in its own past,* it successively solved a formal problem in pattern formation (see Glossary). It, too, proclaims its affiliation under that pattern of patterns which connects.

So far, all the examples that I have offered—the patterns which have membership in the pattern which connects, the anatomy of crab and lobster, the conch, and man and horse—have been superficially static. The examples have been the frozen shapes, results of regularized change, indeed, but themselves finally fixed, like the figures in Keats' "Ode on a Grecian Urn":

> *Fair youth, beneath the trees, thou can'st not leave*
> *Thy song, nor ever can those trees be bare;*
> *Bold lover, never never canst thou kiss,*
> *Though winning near the goal—yet do not grieve;*

She cannot fade, though thou hast not thy bliss,
Forever wilt thou love, and she be fair!

We have been trained to think of patterns, with the exception of those of music, as fixed affairs. It is easier and lazier that way but, of course, all nonsense. In truth, the right way to begin to think about the pattern which connects is to think of it as *primarily* (whatever that means) a dance of interacting parts and only secondarily pegged down by various sorts of physical limits and by those limits which organisms characteristically impose.

There is a story which I have used before and shall use again: A man wanted to know about mind, not in nature, but in his private large computer. He asked it (no doubt in his best Fortran), "Do you compute that you will ever think like a human being?" The machine then set to work to analyze its own computational habits. Finally, the machine printed its answer on a piece of paper, as such machines do. The man ran to get the answer and found, neatly typed, the words:

THAT REMINDS ME OF A STORY

A story is a little knot or complex of that species of connectedness which we call *relevance*. In the 1960s, students were fighting for "relevance," and I would assume that any A is relevant to any B if both A and B are parts or components of the same "story."

Again we face connectedness at more than one level:

First, connection between A and B by virtue of their being components in the same story.

And then, connectedness between people in that all think in terms of stories. (For surely the computer was right. This is indeed how people think.)

Now I want to show that whatever the word *story* means in the story which I told you, the fact of thinking in terms of stories does not isolate human beings as something separate from the starfish and the sea anemones, the coconut palms and the primroses. Rather, if the world be connected, if I am at all fundamentally right in what I am saying, then *thinking in terms of stories* must be shared by all mind or minds, whether ours or those of redwood forests and sea anemones.

Context and relevance must be characteristic not only of all so-called behavior (those stories which are projected out into "action"), but also of all those internal stories, the sequences of the building up of the sea anemone. Its embryology must be somehow made of the stuff of stories. And behind that, again, the evolutionary process through millions of generations whereby the sea anemone, like you and like me, came to be—that process, too, must be of the stuff of stories. There must be relevance in every step of phylogeny and among the steps.

Prospero says, "We are such stuff as dreams are made on," and surely he is nearly right. But I sometimes think that dreams are only fragments of that stuff. It is as if the stuff of which we are made were totally transparent and therefore imperceptible and as if the only appearances of which we can be aware are cracks and planes of fracture in that transparent matrix. Dreams and percepts and stories are perhaps cracks and irregularities in the uniform and timeless matrix. Was this what Plotinus meant by an "invisible and unchanging beauty which pervades all things?"

What is a story that it may connect the As and Bs, its parts? And is it true that the general fact that parts are connected in this way is at the very root of what it is to be alive? I offer you the notion of *context*, of *pattern through time*.

What happens when, for example, I go to a Freudian psychoanalyst? I walk into and create something which we will call a *context* that is at least symbolically (as a piece of the world of ideas) limited and isolated by closing the door. The geography of the room and the door is used as a representation of some strange, nongeographic message.

But I come with stories—not just a supply of stories to deliver to the analyst but stories built into my very being. The patterns and sequences of childhood experience are built into me. Father did so and so; my aunt did such and such; and what they did was outside my skin. But whatever it was that I learned, my learning happened within my experiential sequence of what those important others—my aunt, my father—did.

Now I come to the analyst, this newly important other who must be viewed as a father (or perhaps an antifather) because nothing has meaning except it be seen as in some context. This viewing is called the *transference* and is a general phenomenon in human relations. It is a universal characteristic of all interaction between persons because, after all, the shape of what happened between you and me yesterday carries over to shape how we respond to each other today. And that shaping is, in principle, a *transference* from past learning.

This phenomenon of transference exemplifies the truth of the computer's perception that we think in stories. The analyst must be stretched or shrunk onto the Procrustean bed of the patient's childhood stories. But also, by referring to psychoanalysis, I have narrowed the idea of "story." I have suggested that it has something to do with *context,* a crucial concept, partly undefined and therefore to be examined.

And "context" is linked to another undefined notion called "meaning." Without context, words and actions have no meaning at all. This is true not only of human communication in words but also of all communication whatsoever, of all mental process, of all mind, including that which tells the sea anemone how to grow and the amoeba what he should do next.

I am drawing an analogy between context in the superficial and partly conscious business of personal relations and context in the much deeper, more archaic processes of embryology and homology. I am asserting that whatever the word *context* means, it is an appropriate word, the *necessary* word, in the description of all these distantly related processes.

Let us look at homology backwards. Conventionally, people prove that evolution occurred by citing cases of homology. Let me do the reverse. Let me assume that evolution occurred and go on to ask about the nature of homology. Let us ask what some organ *is* according to the light shed upon it by evolutionary theory.

What is an elephant's trunk? What is it phylogenetically? What did genetics tell it to be?

As you know, the answer is that the elephant's trunk is his "nose." (Even Kipling knew!) And I put the word "nose" in quotation marks because the trunk is being defined by an internal process of communication in growth. The trunk is a "nose" by a process of communication: it is the context of the trunk that identifies it as a nose. That which stands between two eyes and north of a mouth is a "nose," and that is that. It is the *context* that fixes the meaning, and it must surely be the receiving context that provides meaning for the genetic instructions. When I call that a "nose" and this a "hand" I am quoting—or misquoting—the developmental instructions in the growing organism, and quoting what the tissues which received the message thought the message intended.

There are people who would prefer to define noses by their "function"—that of smelling. But if you spell out those definitions, you arrive at the same place using a temporal instead of a spatial context. You attach meaning to the organ by seeing it as playing a given part in sequences of interaction between creature and environment. I call that a *temporal* context. The temporal classification cross-cuts the spatial classification of contexts. But in embryology, the first definition must always be in terms of formal relations. The fetal trunk cannot, in general, smell anything. Embryology is *formal.*

Let me illustrate this species of connection, this connecting pattern, a little further by citing a discovery of Goethe's. He was a considerable botanist who had great ability in recognizing the nontrivial (i.e., in recognizing the patterns that connect). He straightened out the vocabulary of the gross comparative anatomy of flowering plants. He discovered that a "leaf" is not satisfactorily defined as "a flat green thing" or a "stem" as "a cylindrical thing." The way to go about the definition—and undoubtedly somewhere deep in the growth processes of the plant, this is how the matter is handled—is to note that buds (i.e., baby stems) form in the angles of leaves. From that, the botanist constructs the definitions on the basis of the relations between stem, leaf, bud, angle, and so on.

"A stem is that which bears leaves."
"A leaf is that which has a bud in its angle."
"A stem is what was once a bud in that position,"

All that is—or should be—familiar. But the next step is perhaps new.

There is a parallel confusion in the teaching of language that has never been straightened out. Professional linguists nowadays may know what's what, but children in school are still taught nonsense. They are told that a "noun" is the "name of a person, place, or thing," that a "verb" is "an action word," and so on. That is, they are taught at a tender age that the way to define something is by what it supposedly *is* in itself, not by its relation to other things.

Most of us can remember being told that a noun is "the name of a person, place, or thing." And we can remember the utter boredom of parsing or analyzing sentences. Today all that should be changed. Children could be told that a noun is a word having a certain relationship to a predicate. A verb has a certain relation to a noun, its subject. And so on. Relationship could be used as basis for definition, and any child could then see that there is something wrong with the sentence " 'Go' is a verb."

I remember the boredom of analyzing sentences and the boredom later, at Cambridge, of learning comparative anatomy. Both subjects, as taught, were torturously unreal. We *could* have been told something about the pattern which connects: that all communication necessitates context, that without context, there is no meaning, and that contexts confer meaning because there is classification of contexts. The teacher could have argued that growth and differentiation must be controlled by communication. The shapes of animals and plants are transforms of messages. Language is itself a form of communication. The structure of the input must somehow be reflected as structure in the output. Anatomy *must* contain an analogue of grammar because all anatomy is a transform of message material, which must be contextually shaped. And finally, *contextual shaping* is only another term for *grammar*.

So we come back to the patterns of connection and the more abstract, more general (and most empty) proposition that, indeed, there is a pattern of patterns of connection.

This book is built on the opinion that we are parts of a living world. I have placed as epigraph at the head of this chapter a passage from Saint Augustine in which the saint's epistemology is clearly stated. Today such a statement evokes nostalgia. Most of us have lost that sense of unity of biosphere and humanity which would bind and reassure us all with an affirmation of beauty. Most of us do not today believe that whatever the ups and downs of detail within our limited experience, the larger whole is primarily beautiful.

We have lost the core of Christianity. We have lost Shiva, the dancer of Hinduism whose dance at the trivial level is both creation and destruction but in whole is beauty. We have lost Abraxas, the terrible and beautiful god of both day and night in Gnosticism. We have lost totemism, the sense of parallelism between man's organization and that of the animals and plants. We have lost even the Dying God.

We are beginning to play with ideas of ecology, and although we immediately trivialize these ideas into commerce or politics, there is at least an impulse still in the human breast to unify and thereby sanctify the total natural world, of which we are.

Observe, however, that there have been, and still are, in the world many different and even contrasting epistemologies which have been alike in stressing an ultimate unity and, although this is less sure, which have also stressed the notion that ultimate unity is *aesthetic*. The uniformity of these views gives hope that perhaps the great authority of quantitative science may be insufficient to deny an ultimate unifying beauty.

I hold to the presupposition that our loss of the sense of aesthetic unity was, quite simply, an epistemological mistake. I believe that that mistake may be more serious than all the minor insanities that characterize those older epistemologies which agreed upon the fundamental unity.

A part of the story of our loss of the sense of unity has been elegantly told in Lovejoy's *Great Chain of Being,* * which traces the story from classical Greek philosophy to Kant and the beginnings of German idealism in the eighteenth century. This is the story of the idea that the world is/was timelessly created upon *deductive logic*. The idea is clear in the epigraph from *The City of God*. Supreme Mind, or Logos, is at the head of the deductive chain. Below that are the angels, then people, then apes, and so on down to the plants and stones. All is in deductive order and tied into that order by a premise which prefigures our second law of thermodynamics. The premise asserts that the "more perfect" can never be generated by the "less perfect."

In the history of biology, it was Lamarck† who inverted the great chain of being. By insisting that mind is immanent in living crea-

* Arthur O. Lovejoy, *The Great Chain of Being: A Study of the History of an Idea* (Cambridge: Harvard University Press, 1936).

† J.-B. Lamarck, *Philosophie Zoologique* (1809) translated as [Zoological philosophy: An exposition with regard to the natural history of animals, trans. Hugh Elliot] (New York & London: Hafner Press, 1963).

tures and could determine their transformations, he escaped from the negative directional premise that the perfect must always precede the imperfect. He then proposed a theory of "transformism" (which we would call *evolution*) which started from infusoria (protozoa) and marched upward to man and woman.

The Lamarckian biosphere was still a *chain*. The unity of epistemology was retained in spite of a shift in emphasis from transcendent Logos to immanent mind.

The fifty years that followed saw the exponential rise of the Industrial Revolution, the triumph of Engineering over Mind, so that the culturally appropriate epistemology for the *Origin of Species* (1859) was an attempt to exclude mind as an explanatory principle. Tilting at a windmill.

There were protests much more profound than the shrieks of the Fundamentalists. Samuel Butler, Darwin's ablest critic, saw that the denial of mind as an explanatory principle was intolerable and tried to take evolutionary theory back to Lamarckism. But that would not do because of the hypothesis (shared even by Darwin) of the "inheritance of acquired characteristics." This hypothesis—that the responses of an organism to its environment could affect the genetics of the offspring—was an error.

I shall argue that this error was specifically an epistemological error in logical typing and shall offer a definition of *mind* very different from the notions vaguely held by both Darwin and Lamarck. Notably, I shall assume that thought resembles evolution in being a stochastic (see Glossary) process.

In what is offered in this book, the hierarchic structure of thought, which Bertrand Russell called *logical typing,* will take the place of the hierarchic structure of the Great Chain of Being and an attempt will be made to propose a sacred unity of the biosphere that will contain fewer epistemological errors than the versions of that sacred unity which the various religions of history have offered. What is important is that, right or wrong, the epistemology shall be *explicit.* Equally explicit criticism will then be possible.

So the immediate task of this book is to construct a picture of how the world is joined together in its mental aspects. How do ideas, information, steps of logical or pragmatic consistency, and the like fit together? How is logic, the classical procedure for making chains of ideas, related to an outside world of things and creatures, parts and wholes? Do ideas really occur in chains, or is this lineal (see Glossary) structure imposed on them by scholars and philosophers? How is the world of logic, which eschews "circular argument," related to a world in which circular trains of causation are the rule rather than the exception?

What has to be investigated and described is a vast network or matrix of interlocking message material and abstract tautologies, premises, and exemplifications.

But, as of 1979, there is no conventional method of describing such a tangle. We do not know even where to begin.

Fifty years ago, we would have assumed that the best procedures for such a task would have been either logical or quantitative, or both. But we shall see as every schoolboy ought to know that logic is precisely unable to deal with recursive circuits without generating paradox and that quantities are precisely not the stuff of complex communicating systems.

In other words, logic and quantity turn out to be inappropriate devices for describing organisms and their interactions and internal organization. The particular nature of this inappropriateness will be exhibited in due course, but for the moment, the reader is asked to accept as true the assertion that, as of 1979, there is no conventional way of explaining or even describing the phenomena of biological organization and human interaction.

John Von Neumann pointed out thirty years ago, in his *Theory of Games,* that the behavioral sciences lack any reduced model which would do for biology and psychiatry what the Newtonian particle did for physics.

There are, however, a number of somewhat disconnected pieces of wisdom that will aid the task of this book. I shall therefore adopt the method of Little Jack Horner, pulling out plums one after the other and exhibiting them side by side to create an array from which we can go on to list some fundamental criteria of mental process.

In Chapter 2, "Every Schoolboy Knows," I shall gather for the reader some examples of what I regard as simple necessary truths—necessary first if the schoolboy is ever to learn to think and then again necessary because, as I believe, the biological world is geared to these simple propositions.

In Chapter 3 I shall operate in the same way but shall bring to the reader's attention a number of cases in which two or more information sources come together to give information of a sort different from what was in either source separately.

At present, there is no existing science whose special interest is the combining of pieces of information. But I shall argue that the evolutionary process must depend upon such double increments of information. Every evolutionary step is an addition of information to an already existing system. Because this is so, the combinations, harmonies, and discords between successive pieces and layers of information will present many problems of survival and determine many directions of change.

Chapter 4, "The Criteria of Mind," will deal with the characteristics that in fact always seem to be combined in our earthly biosphere to make mind. The remainder of the book will focus more narrowly on problems of biological evolution.

Throughout, the thesis will be that it is possible and worthwhile to *think* about many problems of order and disorder in the biological universe and that we have today a considerable supply of tools of thought which we do not use, partly because—professors and schoolboys alike—we are ignorant of many currently available insights and partly because we are unwilling to accept the necessities that follow from a clear view of the human dilemmas.

II

EVERY
SCHOOLBOY
KNOWS...

By education most have been misled;
So they believe, because they so were bred.
The priest continues what the nurse began,
And thus the child imposes on the man.
—JOHN DRYDEN, *The Hind and the Panther*

Science, like art, religion, commerce, warfare, and even sleep, is based on *presuppositions*. It differs, however from most other branches of human activity in that not only are the pathways of scientific thought determined by the presuppositions of the scientists but their goals are the testing and revision of old presuppositions and the creation of new.

In this latter activity, it is clearly desirable (but not absolutely necessary) for the scientist to know consciously and be able to state his own presuppositions. It is also convenient and necessary for scientific judgment to know the presuppositions of colleagues working in the same field. Above all, it is necessary for the reader of scientific matter to know the presuppositions of the writer.

I have taught various branches of behavioral biology and cultural anthropology to American students, ranging from college freshmen to psychiatric residents, in various schools and teaching hospitals, and I have encountered a very strange gap in their thinking that springs from a lack of certain *tools* of thought. This lack is rather equally distributed at all levels of education, among students of both sexes and among humanists as well as scientists. Specifically, it is lack of knowledge of the presuppositions not only of science but also of everyday life.

This gap is, strangely, less conspicuous in two groups of students that might have been expected to contrast strongly with each other: the Catholics and the Marxists. Both groups have thought about or have been told a little about the last 2,500 years of human thought, and both groups have some recognition of the importance of philosophic, scientific, and epistemological presuppositions. Both groups are difficult to teach because they attach such great importance to "right" premises and presuppositions that heresy becomes for them a threat of excommunication. Naturally, anybody who feels heresy to be a danger will devote some care to being conscious of his or her own presuppositions and will develop a sort of connoisseurship in these matters.

Those who lack all idea that it is possible to be wrong can learn nothing except know-how.

The subject matter of this book is notably close to the core of religion and to the core of scientific orthodoxy. The presuppositions— and most students need some instruction in what a presupposition looks like—are matters to be brought out into the open.

There is, however, another difficulty, almost peculiar to the American scene. Americans are, no doubt, as rigid in their presuppositions as any other people (and as rigid in these matters as the writer of this book), but they have a strange response to any articulate statement of presupposition. Such statement is commonly assumed to be hostile or mocking or—and this is the most serious—is heard to be *authoritarian*.

It thus happens that in this land founded for the freedom of religion, the teaching of religion is outlawed in the state educational system. Members of weakly religious families get, of course, no religious training from any source outside the family.

Consequently, to make any statement of premise or presupposition in a formal and articulate way is to challenge the rather subtle resistance, not of contradiction, because the hearers do not know the contradictory premises nor how to state them, but of the cultivated deafness that children use to keep out the pronouncements of parents, teachers, and religious authorities.

Be all that as it may, I believe in the importance of scientific presuppositions, in the notion that there are better and worse ways of constructing scientific theories, and in insisting on the articulate statement of presuppositions so that they may be improved.

Therefore, this chapter is devoted to a list of presuppositions, some familiar, some strange to readers whose thinking has been protected from the harsh notion that some propositions are simply wrong. Some tools of thought are so blunt that they are almost useless; others are so sharp that they are dangerous. But the wise man will have the use of both kinds.

It is worthwhile to attempt a tentative recognition of certain basic presuppositions which all *minds* must share or, conversely, to define mind by listing a number of such basic communicational characteristics.

1. SCIENCE NEVER PROVES ANYTHING

Science sometimes *improves* hypotheses and sometimes *disproves* them. But *proof* would be another matter and perhaps never occurs except in the realms of totally abstract tautology. We can sometimes say that *if* such and such abstract suppositions or postulates are given, *then* such and such must follow absolutely. But the truth about what can be *perceived* or arrived at by induction from perception is something else again.

Let us say that truth would mean a precise correspondence between our description and what we describe or between our total network of abstractions and deductions and some total understanding of the outside world. Truth in this sense is not obtainable. And even if we ignore the barriers of coding, the circumstance that our description will be in words or figures or pictures but that what we describe is going to be in flesh and blood and action—even disregarding that hurdle of translation, we shall never be able to claim final knowledge of anything whatsoever.

A conventional way of arguing this matter is somewhat as follows: Let us say that I offer you a series—perhaps of numbers, perhaps of other indications—and that I provide the presupposition that the series is ordered. For the sake of simplicity, let it be a series of numbers:

2, 4, 6, 8, 10, 12

Then I ask you, "What is the next number in this series?" You will probably say, "14."

But if you do, I will say, "Oh, no. The next number is 27." In other words, the generalization to which you jumped from the data given in the first instance—that the series was the series of even numbers—was proved to be wrong or only approximate by the next event.

Let us pursue the matter further. Let me continue my statement by creating a series as follows:

2, 4, 6, 8, 10, 12, 27, 2, 4, 6, 8, 10, 12, 27, 2, 4, 6, 8, 10, 12, 27 . . .

Now if I ask you to guess the next number, you will probably say, "2." After all, you have been given three repetitions of the sequence from 2 to 27; and if you are a good scientist, you will be influenced by the presupposition called *Occam's razor,* or the *rule of parsimony:* that is, a preference for the simplest assumptions that will fit the facts. On the basis of simplicity, you will make the next prediction. But those facts—what are they? They are not, after all, available to you beyond the end of the (possibly incomplete) sequence that has been given.

You *assume* that you can predict, and indeed I suggested this presupposition to you. But the only basis you have is your (trained) preference for the simpler answer and your trust that my challenge indeed meant that the sequence was incomplete and ordered.

Unfortunately (or perhaps fortunately), it is so that the next fact is never available. All you have is the hope of simplicity, and the next fact may always drive you to the next level of complexity.

Or let us say that for any sequence of numbers I can offer, there will always be a few ways of describing that sequence which will be simple, but there will be an *infinite* number of alternative ways not limited by the criterion of simplicity.

Suppose the numbers are represented by letters:

x, w, p, n

and so on. Such letters could stand for any numbers whatsoever, even fractions. I have only to repeat the series three or four times in some verbal or visual or other sensory form, even in the forms of pain or kinesthesia, and you will begin to perceive pattern in what I offer you. It will become in your mind—and in mine—a theme, and it will have aesthetic value. To that extent, it will be familiar and understandable.

But the pattern may be changed or broken by addition, by repetition, by anything that will force you to a new perception of it, and these changes can never be predicted with absolute certainty because they have not yet happened.

We do not know enough about how the present will lead into the future. We shall never be able to say, "Ha! My perception, my accounting for that series, will indeed cover its next and future components," or "Next time I meet with these phenomena, I shall be able to predict their total course."

Prediction can never be absolutely valid and therefore science can never *prove* some generalization or even *test* a single descriptive statement and in that way arrive at final truth.

There are other ways of arguing this impossibility. The argument of this book—which again, surely, can only convince you insofar as what I say fits with what you know and which may be collapsed or totally changed in a few years—presupposes that science is a *way of perceiving* and making what we may call "sense" of our percepts. But perception operates only upon difference. All receipt of information is necessarily the receipt of news of *difference,* and all perception of difference is limited by threshold. Differences that are too slight or too slowly presented are not perceivable. They are not food for perception.

It follows that what we, as scientists, can perceive is always limited by threshold. That is, what is subliminal will not be grist for our mill. Knowledge at any given moment will be a function of the thresholds of our available means of perception. The invention of the microscope or the telescope or of means of measuring time to the fraction of a nanosecond or weighing quantities of matter to millionths of a gram— all such improved devices of perception will disclose what was utterly unpredictable from the levels of perception that we could achieve before that discovery.

Not only can we not predict into the next instant of the future, but, more profoundly, we cannot predict into the next dimension of the microscopic, the astronomically distant, or the geologically ancient. As a method of perception—and that is all science can claim to be—science, like all other methods of perception, is limited in its ability to collect the outward and visible signs of whatever may be truth.

Science *probes;* it does not prove.

2. THE MAP IS NOT THE TERRITORY, AND THE NAME IS NOT THE THING NAMED

This principle, made famous by Alfred Korzybski, strikes at many levels. It reminds us in a general way that when we think of coconuts or pigs, there are no coconuts or pigs in the brain. But in a more abstract way, Korzybski's statement asserts that in all thought or perception or communication about perception, there is a transformation, a coding, between the report and the thing reported, the *Ding an sich.* Above all, the relation between the report and that mysterious thing reported tends to have the nature of a *classification,* an assignment of the thing to a class. Naming is always classifying, and mapping is essentially the same as naming.

Korzybski was, on the whole, speaking as a philosopher, attempting to persuade people to discipline their manner of thinking. But he could not win. When we come to apply his dictum to the natural history of human mental process, the matter is not quite so simple. The distinction between the name and the thing named or the map and the territory is perhaps really made only by the dominant hemisphere of the brain. The symbolic and affective hemisphere, normally on the right-hand side, is probably unable to distinguish name from thing named. It is certainly not concerned with this sort of distinction. It therefore happens that certain nonrational types of behavior are necessarily present in human life. We do, in fact, have two hemispheres; and we cannot get away from that fact. Each hemisphere does, in fact, operate somewhat differently from the other, and we cannot get away from the tangles that that difference proposes.

For example, with the dominant hemisphere, we can regard such a thing as a flag as a sort of name of the country or organization that it represents. But the right hemisphere does not draw this distinction and regards the flag as sacramentally identical with what it represents. So "Old Glory" is the United States. If somebody steps on it, the response may be rage. And this rage will not be diminished by an explanation of map-territory relations. (After all, the man who tramples the flag is equally identifying it with that for which it stands.) There will always and necessarily be a large number of situations in which the response is not guided by the logical distinction between the name and the thing named.

3. THERE IS NO OBJECTIVE EXPERIENCE

All experience is subjective. This is only a simple corollary of a point made in section 4: that our brains make the images that we think we "perceive."

It is significant that all perception—all conscious perception—has image characteristics. A pain is localized somewhere. It has a beginning and an end and a location and stands out against a background. These are the elementary components of an image. When somebody steps on my toe, what I experience is, not his stepping on my toe, but my *image* of his stepping on my toe reconstructed from neural reports reaching my brain somewhat after his foot has landed on mine. Experience of the exterior is always mediated by particular sense organs and neural pathways. To that extent, objects are my creation, and my experience of them is subjective, not objective.

It is, however, not a trivial assertion to note that very few persons, at least in occidental culture, doubt the objectivity of such sense data as pain or their visual images of the external world. Our civilization is deeply based on this illusion.

4. THE PROCESSES OF IMAGE FORMATION
ARE UNCONSCIOUS

This generalization seems to be true of everything that happens between my sometimes conscious action of directing a sense organ at some source of information and my conscious action of deriving information from an image that "I" seem to see, hear, feel, taste, or smell. Even a pain is surely a created image.

No doubt men and donkeys and dogs are all conscious of listening and even of cocking their ears in the direction of sound. As for sight, something moving in the periphery of my visual field will call "attention" (whatever that means) so that I shift my eyes and even my head to look at it. This is often a conscious act, but it is sometimes so nearly automatic that it goes unnoticed. Often I am conscious of turning my head but unaware of the peripheral sighting that caused me to turn. The peripheral retina receives a lot of information that remains outside consciousness—possibly but not certainly in image form.

The *processes* of perception are inaccessible; only the *products* are conscious and, of course, it is the products that are necessary. The two general facts—first, that I am unconscious of the process of making the images which I consciously see and, second, that in these unconscious processes, I use a whole range of presuppositions which become built into the finished image—are, for me, the beginning of empirical epistemology.

Of course, we all know that the images which we "see" are indeed manufactured by the brain or mind. But to know this in an intellectual sense is very different from realizing that it is truly so. This aspect of the matter came forcibly to my attention some thirty years ago in New York, where Adalbert Ames, Jr., was demonstrating his experiments on how we endow our visual images with depth. Ames was an ophthalmologist who had worked with patients who suffered from anisoconia; that is, they formed images of different sizes in the two eyes. This led him to study the subjective components of the perception of depth. Because this matter is important and provides the very basis of empirical or experimental epistemology, I will narrate my encounter with the Ames experiments in some detail.

Ames had the experiments set up in a large, empty apartment in

New York City. There were, as I recall, some fifty experiments. When I arrived to see the show, I was the only visitor. Ames greeted me and suggested that I start at the beginning of the sequence of demonstrations while he went back to work for awhile in a small room furnished as an office. Otherwise, the apartment contained no furniture except for two folding deck chairs.

I went from one experiment to the next. Each contained some sort of optical illusion affecting the perception of depth. The thesis of the whole series was that we use five main clues to guide us in creating the appearance of depth in the images that we create as we look out through our eyes at the world.

The first of these clues is size;* that is, the size of the physical image on the retina. Of course, we cannot *see* this image so it would be more exact to say that the first clue to distance is the angle which the object subtends at the eye. But indeed this angle is also not visible. The clue to distance which is reported on the optic nerve is perhaps *change in angle subtended.*† The demonstration of this truth was a pair of balloons in a dark area. The balloons themselves were equally illuminated, but their air could be passed from one balloon into the other. The balloons themselves did not move, but as one grew and the other shrank, it appeared to the observer that the one which grew, approached, and the one which shrank, retreated. As the air was shifted from one balloon to the other and back again, the balloons appeared to move alternately forward and back.

The second clue was contrast in brightness. To demonstrate this, the balloons stayed the same size and, of course, did not really move. Only the illumination changed, shining first on one balloon and then on the other. This alternation of illumination, like the alternation in size, gave the balloons the appearance of approaching and retreating in turn as the light fell first on one and then on the other.

Then the sequence of experiments showed that these two clues, size and brightness, could be played against each other to give a contradiction. The shrinking balloon now always got the more light. This

---

*More precisely, I should have written: "The first of these clues is *contrast* in size . . ."
†I observe not only that the processes of visual perception are inaccessible to consciousness but also that it is impossible to construct in words any acceptable description of what must happen in the simplest act of seeing. For that which is not conscious, the language provides no means of expression.

combined experiment introduced the idea that some clues are dominant over others.

The total sequence of clues demonstrated that day included size, brightness, overlap, binocular parallax, and parallax created by movements of the head. Of these, the most strongly dominant was parallax by head motion.

After looking at twenty or thirty such demonstrations, I was ready to take a break and went to sit in one of the folding deck chairs. It collapsed under me. Hearing the noise, Ames came out to check that all was well. He then stayed with me and demonstrated the two following experiments.

The first dealt with parallax (see Glossary). On a table perhaps five feet long, there were two objects: a pack of Lucky Strike cigarettes, supported on a slender spike some inches from the surface of the table and a book of paper matches, similarly raised on a spike, at the far end of the table.

Ames had me stand at the near end of the table and describe what I saw; that is, the location of the two objects and how big they seemed to be. (In Ames's experiments, you are always made to observe the truth before being subjected to the illusions.)

Ames then pointed out to me that there was a wooden plank with a plain round hole in it set upright at the edge of the table at my end so that I could look through the hole down the length of the table. He had me look through this hole and tell him what I saw. Of course, the two objects still appeared to be where I knew them to be and to be of their familiar sizes.

Looking through the hole in the plank, I had lost the crow's-eye view of the table and was reduced to the use of a single eye. But Ames suggested that I could get parallax on the objects by sliding the plank sideways.

As I moved my eye sideways with the plank, the image changed totally—as if by magic. The Lucky Strike pack was suddenly at the far end of the table and appeared to be about twice as tall and twice as wide as a normal pack of cigarettes. Even the surface of the paper of which the pack was made had changed in texture. Its small irregularities were now seemingly larger. The book of matches, on the other hand, suddenly appeared to be of dollhouse size and to be located halfway down the length of the table in the position where the pack of cigarettes had formerly been seen to be.

What had happened?

The answer was simple. Under the table, where I could not see them, there were two levers or rods that moved the two objects sideways as I moved the plank. In normal parallax, as we all know, when we look out from a moving train, the objects close to us appear to be left behind fast; the cows beside the railroad track do not stay to be observed. The distant mountains, on the other hand, are left behind so slowly that, in contrast with the cows, they seem almost to travel with the train.

In this case, the levers under the table caused the nearer object to move along with the observer. The cigarette pack was made to act as if it were far away; the book of matches was made to move as if it were close by.

In other words, by moving my eye and with it the plank, I created a reversed appearance. Under such circumstances, the unconscious processes of image formation made the appropriate image. The information from the cigarette pack was read and built up to be the image of a distant pack, but the height of the pack still subtended the same angle at the eye. Therefore, the pack now appeared to be of giant size. The book of matches, correspondingly, was brought seemingly close but still subtended the same angle that it subtended from its true location. What I created was an image in which the book of matches appeared to be half as far away and half its familiar size.

The machinery of perception created the image in accordance with the rules of parallax, rules that were for the first time clearly verbalized by painters in the Renaissance; and this whole process, the creating of the image with its built-in conclusions from the clues of parallax, happened quite outside my consciousness. The rules of the universe that we think we know are deep buried in our processes of perception.

Epistemology, at the natural history level, is mostly unconscious and correspondingly difficult to change. The second experiment that Ames demonstrated illustrates this difficulty of change.

This experiment has been called the *trapezoidal room*. In this case, Ames had me inspect a large box about five feet long, three feet high, and three feet deep from front to back. The box was of strange trapezoidal shape, and Ames asked me to examine it carefully in order to learn its true shape and dimensions.

In the front of the box was a peephole big enough for two eyes, but before beginning the experiment, Ames had me put on a pair of prismatic spectacles that would corrupt my binocular vision. I was to have the subjective presupposition that I had the parallax of two eyes when indeed I had almost no binocular clues.

When I looked in through the peephole, the interior of the box appeared to be quite rectangular and was marked out like a room with rectangular windows. The true lines of paint suggesting windows were, of course, far from simple; they were drawn to give the impression of rectangularity, contradicting the true trapezoidal shape of the room. The side of the box toward which I faced when looking through the peephole was, I knew from my earlier inspection, obliquely placed, so that it was further from me at the left end and closer to me on the right.

Ames gave me a stick and asked me to reach in and touch with the point of the stick a sheet of typewriting paper pinned to the left-hand wall. I managed this fairly easily. Ames then said, "Do you see a similar piece of paper on the right-hand side? I want you to hit that second piece of paper with the stick. Start with the end of your stick against the left-hand paper, and hit as hard as you can."

I smote hard. The end of my stick moved about an inch and then hit the back of the room and could move no farther. Ames said, "Try again."

I tried perhaps fifty times, and my arm began to ache. I knew, of course, what correction I had to impose on my movement: I had to pull in as I struck in order to avoid that back wall. But what I *did* was governed by my image. I was trying to pull against my own spontaneous movement. (I suppose that if I had shut my eyes, I could have done better, but I did not try that.)

I never did succeed in hitting the second piece of paper, but, interestingly, my performance improved. I was finally able to move my stick several inches before it hit the back wall. And *as I practiced and improved my action,* my image changed to give me a more trapezoidal impression of the room's shape.

Ames told me afterward that, indeed, with more practice, people learned to hit the second paper very easily and, at the same time, learned to see the room in its true trapezoidal shape.

The trapezoidal room was the last in the sequence of experiments, and after it, Ames suggested that we go to lunch. I went to wash up in the bathroom of the apartment. I turned the faucet marked "C" and got a jet of boiling water mixed with steam.

Ames and I then went down to find a restaurant. My faith in my own image formation was so shaken that I could scarcely cross the street. I was not sure that the oncoming cars were really where they seemed to be from moment to moment.

In sum, there is no free will against the immediate commands of the images that perception presents to the "mind's eye." But through arduous practice and self-correction, it is partly possible to alter those images. (Such changes in *calibration* are further discussed in Chapter 7.)

In spite of this beautiful experimentation, the fact of image formation remains almost totally mysterious. How it is done, we know not—nor, indeed, for what purpose.

It is all very well to say that it makes a sort of adaptive sense to present only the images to consciousness without wasting psychological process on consciousness of their making. But there is no clear primary reason for using images at all or, indeed, for being *aware* of any part of our mental processes.

Speculation suggests that image formation is perhaps a convenient or economical method of passing information across some sort of *interface*. Notably, where a person must act in a context between two machines, it is convenient to have the machines feed their information to him or her in image form.

A case that has been studied systematically is that of a gunner controlling antiaircraft fire on a naval ship.* The information from a series of sighting devices aimed at a flying target is summarized for the gunner in the form of a moving dot on a screen (i.e., an image). On the same screen is a second dot, whose position summarizes the direction in which an antiaircraft gun is aimed. The man can move this second dot

* John Stroud, personal communication.

by turning knobs on the device. These knobs also change the gun's aim. The man must operate the knobs until the dots coincide on the screen. He then fires the gun.

The system contains two interfaces: sensory system—man and man—effector system. Of course, it is conceivable that in such a case, both the input information and the output information could be processed in digital form, without transformation into an iconic mode. But it seems to me that the iconic device is surely more convenient not only because, being human, I am a maker of mental images but also because at these interfaces images are economical or efficient. If that speculation is correct, then it would be reasonable to guess that mammals form images because the mental processes of mammals must deal with many interfaces.

There are some interesting side effects of our unawareness of the processes of perception. For example, when these processes work unchecked by input material from a sense organ, as in dream or hallucination or eidetic (see Glossary) imagery, it is sometimes difficult to doubt the external reality of what the images seem to represent. Conversely, it is perhaps a very good thing that we do *not* know too much about the work of creating perceptual images. In our ignorance of that work, we are free to *believe* what our senses tell us. To doubt continually the evidence of sensory report might be awkward.

5. THE DIVISION OF THE PERCEIVED UNIVERSE INTO PARTS AND WHOLES IS CONVENIENT AND MAY BE NECESSARY,* BUT NO NECESSITY DETERMINES HOW IT SHALL BE DONE

I have tried many times to teach this generality to classes of students and for this purpose have used Figure 1. The figure is presented to the class as a reasonably accurate chalk drawing on the blackboard, but without the letters marking the various angles. The class is asked to

* The question of formal necessity raised here might have an answer as follows: Evidently, the universe is characterized by an uneven distribution of causal and other types of linkage between its parts; that is, there are regions of dense linkage separated from each other by regions of less dense linkage. It may be that there are necessarily and inevitably processes which are responsive to the density of interconnection so that density is increased or sparsity is made more sparse. In such a case, the universe would necessarily present an appearance in which wholes would be bounded by the relative sparseness of their interconnection.

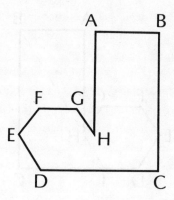

Figure 1

describe "it" in a page of written English. When each student has finished his or her description, we compare the results. They fall into several categories:

    a. About 10 percent or less of students say, for example, that
the object is a boot or, more picturesquely, the boot of a man with a
gouty toe or even a toilet. Evidently, from this and similar analogic or
iconic descriptions, it would be difficult for the hearer of the description
to reproduce the object.

    b. A much larger number of students see that the object con-
tains most of a rectangle and most of a hexagon, and having divided it
into parts in this way, they then devote themselves to trying to describe
the relations between the incomplete rectangle and hexagon. A small
number of these (but, surprisingly, usually one or two in every class)
discover that a line, *BH,* can be drawn and extended to cut the base
line, *DC,* at a point *I* in such a way that *HI* will complete a regular
hexagon (Figure 2). This imaginary line will define the proportions of
the rectangle but not, of course, the absolute lengths. I usually congrat-
ulate these students on their ability to create what resembles many scientific hypotheses, which "explain" a perceptible regularity in terms of
some entity created by the imagination.

    c. Many well-trained students resort to an operational method of
description. They will start from some point on the outline of the object
(interestingly enough, always an angle) and proceed from there, usually
clockwise, with instructions for drawing the object.

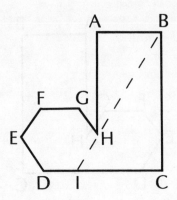

Figure 2

d. There are also two other well-known ways of description that no student has yet followed. No student has started from the statement "It's made of chalk and blackboard." No student has ever used the method of the halftone block, dividing the surface of the blackboard into a grid (arbitrarily rectangular) and reporting "yes" and "no" on whether each box of the grid contains or does not contain some part of the object. Of course, if the grid is coarse and the object small, a very large amount of information will be lost. (Imagine the case in which the entire object is smaller than the grid unit. The description will then consist of not more than four nor less than one affirmation, according to how the divisions of the grid fall upon the object.) However, this is, in principle, how the halftone blocks of newspaper illustration are transmitted by electric impulse and, indeed, how television works.

Note that all these methods of description contribute nothing to an *explanation* of the object—the hexago-rectangle. Explanation must always grow out of description, but the description from which it grows will always necessarily contain arbitrary characteristics such as those exemplified here.

6. DIVERGENT SEQUENCES ARE UNPREDICTABLE

According to the popular image of science, everything is, in principle, predictable and controllable; and if some event or process is not predictable and controllable in the present state of our knowledge, a little more knowledge and, especially, a little more know-how will enable us to predict and control the wild variables.

This view is wrong, not merely in detail, but in principle. It is even possible to define large classes of phenomena where prediction and control are simply impossible for very basic but quite understandable reasons. Perhaps the most familiar example of this class of phenomena is the breaking of any superficially homogeneous material, such as glass. The Brownian movement (see Glossary) of molecules in liquids and gases is similarly unpredictable.

If I throw a stone at a glass window, I shall, under appropriate circumstances, break or crack the glass in a star-shaped pattern. If my stone hits the glass as fast as a bullet, it is possible that it will detach from the glass a neat conical plug called a *cone of percussion.* If my stone is too slow and too small, I may fail to break the glass at all. Prediction and control will be quite possible at this level. I can easily make sure which of three results (the star, the percussion cone, or no breakage) I shall achieve, provided I avoid marginal strengths of throw.

But within the conditions which produce the star-shaped break, it will be impossible to predict or control the pathways and the positions of the arms of the star.

Curiously enough, the more precise my laboratory methods, the more unpredictable the events will become. If I use the most homogeneous glass available, polish its surface to the most exact optical flatness, and control the motion of my stone as precisely as possible, ensuring an almost precisely vertical impact on the surface of the glass, all my efforts will only make the events more impossible to predict.

If, on the other hand, I scratch the surface of the glass or use a piece of glass that is already cracked (which would be cheating), I shall be able to make some approximate predictions. For some reason (unknown to me), the break in the glass will run parallel to the scratch and about 1/100 of an inch to the side, so that the scratch mark will appear on only one side of the break. Beyond the end of the scratch, the break will veer off unpredictably.

Under tension, a chain will break at its weakest link. That much is predictable. What is difficult is to identify the weakest link before it breaks. *The generic we can know, but the specific eludes us.* Some chains are designed to break at a certain tension and at a certain link. But a good chain is homogeneous, and no prediction is possible. And because we cannot know which link is weakest, we cannot know precisely how much tension will be needed to break the chain.

If we heat a clear liquid (say, clean distilled water) in a clean, smooth beaker, at what point will the first bubble of steam appear? At what temperature? And at what instant?

These questions are unanswerable unless there is a tiny roughness in the inner surface of the beaker or a speck of dust in the liquid. In the absence of such an evident nucleus for the beginning of the change of state, no prediction is possible; and because we cannot say where the change will start, we also cannot say *when.* Therefore, we cannot say at what temperature boiling will begin.

If the experiment is critically performed—that is, if the water is very clean and the beaker very smooth—there will be some superheating. In the end, the water will boil. In the end, there will always be a *difference* that can serve as the nucleus for the change. In the end, the superheated liquid will "find" this differentiated spot and will boil explosively for a few moments until the temperature is reduced to the regular boiling point appropriate to the surrounding barometric pressure.

The freezing of liquid is similar, as is the falling out of crystals from a supersaturated solution. A nucleus—that is, a differentiated point, which in the case of a supersaturated solution may, indeed, be a microscopic crystal—is needed for the process to start.

We shall note elsewhere in this book that there is a deep gulf between statements about an identified individual and statements about a class. Such statements are of *different logical type,* and prediction from one to the other is always unsure. The statement "The liquid is boiling" is of different logical type from the statement "That molecule will be the first to go."

This matter has a number of sorts of relevance to the theory of history, to the philosophy behind evolutionary theory, and in general, to our understanding of the world in which we live.

In the theory of history, Marxian philosophy, following Tolstoi, insists that the great men who have been the historic nuclei for profound social change or invention are, in a certain sense, irrelevant to the changes they precipitated. It is argued, for example, that in 1859, the occidental world was ready and ripe (perhaps overripe) to create and receive a theory of evolution that could reflect and justify the ethics of the Industrial Revolution. From that point of view, Charles Darwin himself could be made to appear unimportant. If he had not put out his theory, somebody else would have put out a similar theory within the next five years. Indeed, the parallelism between Alfred Russel Wallace's theory and that of Darwin would seem at first sight to support this view.*

The Marxians would, as I understand it, argue that there is bound to be a weakest link, that under appropriate social forces† or tensions, some individual will be the first to start the trend, and that it does not matter who.

But, of course, it *does* matter who starts the trend. If it had been Wallace instead of Darwin, we would have had a very different theory of evolution today. The whole cybernetics movement might have occurred 100 years earlier as a result of Wallace's comparison between the steam engine with a governor and the process of natural selection. Or perhaps the big theoretical step might have occurred in France and evolved from the ideas of Claude Bernard who in the late nineteenth century, discovered what later came to be called the *homeostasis* of the body. He observed that the *milieu interne*—the internal environment—was balanced, or self-correcting.

It is, I claim, nonsense to say that it does not matter which individual man acted as the nucleus for the change. *It is precisely this that makes history unpredictable into the future.* The Marxian error is a simple blunder in logical typing, a confusion of individual with class.

---

* The story is worth repeating. Wallace was a young naturalist who, in 1856 (three years before the publication of Darwin's *Origin*), while in the rain forests of Ternate, Indonesia, had an attack of malaria and, following delirium, a psychedelic experience in which he discovered the principle of natural selection. He wrote this out in a long letter to Darwin. In this letter he explained his discovery in the following words: "The action of this principle is exactly like that of the centrifugal governor of the steam engine, which checks and corrects any irregularities almost before they become evident; and in like manner no unbalanced deficiency in the animal kingdom can ever reach any conspicuous magnitude because it would make itself felt at the very first step, by rendering existence difficult and extinction almost sure to follow." (Reprinted in *Darwin, a Norton Critical Edition,* ed. Philip Appleman, W. W. Norton, 1970.)

† Notice the use of physical metaphor, inappropriate to the creatural phenomena being discussed. Indeed, it may be argued that this whole comparison between social biological matters, on the one hand, and physical processes, on the other, is a monstrous use of inappropriate metaphor.

7. CONVERGENT SEQUENCES ARE PREDICTABLE

This generality is the converse of the generality examined in section 6, and the relation between the two depends on the contrast between the concepts of divergence and convergence. This contrast is a special case, although a very fundamental one, of the difference between successive levels in a Russellian hierarchy, a matter to be discussed in Chapter 4. For the moment, it should be noted that the components of a Russellian hierarchy are to each other as member to class, as class to class of classes, or as thing named to name.

What is important about divergent sequences is that our description of them concerns *individuals,* especially individual molecules. The crack in the glass, the first step in the beginning of the boiling of water, and all the rest are cases in which the location and instant of the event is determined by some momentary constellation of a small number of individual molecules. Similarly, any description of the pathways of individual molecules in Brownian movement allows for no extrapolation. What happens at one moment, even if we could know it, would not give us data to predict what will happen at the next.

In contrast, the movement of planets in the solar system, the trend of a chemical reaction in an ionic mixture of salts, the impact of billiard balls, which involves millions of molecules—all are predictable because our description of the events has as its subject matter the behavior of immense crowds or classes of individuals. It is this that gives science some justification for statistics, providing the statistician always remembers that his statements have reference only to aggregates.

In this sense, the so-called laws of probability mediate between descriptions of the behavior of the individual and descriptions of that of the gross crowd. We shall see later that this particular sort of conflict between the individual and the statistical has dogged the development of evolutionary theory from the time of Lamarck onward. If Lamarck had asserted that changes in environment would affect the general characteristics of whole populations, he would have been in step with the latest genetic experiments such as those of Waddington on genetic assimilation, to be discussed in Chapter 6. But Lamarck and, indeed, his followers ever since have seemed to have an innate proclivity for confusion of logical types. (This matter and the corresponding confusions of ortho-dox evolutionists will be discussed in Chapter 6.)

Be all that as it may, in the stochastic processes (see Glossary) either of evolution or of thought, the new can be plucked from nowhere but the random. And to pluck the new from the random, if and when it happens to show itself, requires some sort of selective machinery to account for the ongoing persistence of the new idea. Something like *natural selection,* in all its truism and tautology, must obtain. To persist, the new must be of such a sort that it will endure longer than the alternatives. What lasts longer among the ripples of the random must last longer than those ripples that last not so long. That is the theory of natural selection in a nutshell.

The Marxian view of history—which in its crudest form would argue that if Darwin had not written *The Origin of Species,* somebody else would have produced a similar book within the next five years—is an unfortunate effort to apply a theory that would view social process as *convergent* to events involving unique human beings. The error is, again, one of logical typing.

8. "NOTHING WILL COME OF NOTHING"

This quotation from *King Lear* telescopes into a single utterance a whole series of medieval and more modern wise saws. These include:

a. The law of the conservation of matter and its converse, that no new matter can be expected to make an appearance in the laboratory. (Lucretius said, "Nothing can ever be created out of nothing by divine power." *)

b. The law of the conservation of energy and its converse, that no new energy can be expected in the laboratory.

c. The principle demonstrated by Pasteur, that no new living matter can be expected to appear in the laboratory.

d. The principle that no new order or pattern can be created without *information*.

* Lucretius, *On the Nature of the Universe,* translated by Ronald E. Lathan (Baltimore: Penguin Books).

Of all these and other similar negative statements, it may be said that they are rules for expectation rather than laws of nature. They are so nearly true that all exceptions are of extreme interest.

What is especially interesting is hidden in the relations between these profound negations. For example, we know today that between the conservation of energy and the conservation of matter, there is a bridge whereby each of these negations is itself negated by an interchange of matter into energy and, presumably, of energy into matter.

In the present connection, however, it is the last of the series that is of chief interest, the proposition that in the realms of communication, organization, thought, learning, and evolution, "nothing will come of nothing" without *information*.

This law differs from the conservative laws of energy and mass in that it contains no clause to deny the destruction and loss of information, pattern, or negative entropy. Alas—but also be glad of it—pattern and/or information is all too easily eaten up by the random. The messages and guidelines for order exist only, as it were, in sand or are written on the surface of waters. Almost any disturbance, even mere Brownian movement, will destroy them. Information can be forgotten or blurred. The code books can be lost.

The messages cease to be messages when nobody can read them. Without a Rosetta stone, we would know nothing of all that was written in Egyptian hieroglyphs. They would be only elegant ornaments on papyrus or rock. To be meaningful—even to be recognized as pattern— every regularity must meet with complementary regularities, perhaps skills, and these skills are as evanescent as the patterns themselves. They, too, are written on sand or the surface of waters.

The genesis of the skill to respond to the message is the obverse, the other side of the process of evolution. It is *coevolution* (see Glossary).

Paradoxically, the deep partial truth that "nothing will come of nothing" in the world of information and organization encounters an interesting contradiction in the circumstance that *zero*, the complete absence of any indicative event, can be a message. The larval tick climbs a tree and waits on some outer twig. If he smells sweat, he falls, perhaps landing on a mammal. But if he smells *no sweat* after some weeks, he falls and goes to climb another tree.

The letter that you do not write, the apology you do not offer, the food that you do not put out for the cat—all these can be sufficient and effective messages because zero, *in context,* can be meaningful; and it is the recipient of the message who creates the context. This power to create *context* is the recipient's skill; to acquire which is his half of the coevolution mentioned above. He or she must acquire that skill by learning or by lucky mutation, that is, by a successful raid on the random. The recipient must be, in some sense, ready for the appropriate discovery when it comes.

Thus, the converse of the proposition that "nothing will come of nothing" without information is conceivably possible with stochastic process. *Readiness* can serve to select components of the random which thereby become new information. But always a supply of random appearances must be available from which new information can be made.

This circumstance splits the entire field of organization, evolution, maturation and learning, into two separate realms, of which one is the realm of epigenesis, or embryology, and the other the realm of evolution and learning.

Epigenesis is the word preferred by C. H. Waddington for his central field of interest, whose old name was *embryology.* It stresses the fact that every embryological step is an act of *becoming* (Greek *genesis*) which must be built *upon* (Greek *epi*) the immediate status quo ante. Characteristically, Waddington was contemptuous of conventional information theory, which allowed nothing, as he saw it, for the "new" information he felt was generated at each stage of epigenesis. Indeed, according to conventional theory, there is no new information in this case.

Ideally, epigenesis should resemble the development of a complex tautology (see Glossary) in which nothing is added after the axioms and definitions have been laid down. The Pythagorean theorem is implicit (i.e., already folded into) Euclid's axioms, definitions, and postulates. All that is required is its unfolding and, for human beings, some knowledge of the order of steps to be taken. This latter species of information will become necessary only when Euclid's tautology is modeled in words and symbols sequentially arranged on paper or in time. In the ideal tautology, there is no time, no unfolding, and no argument. What is implicit is there, but, of course, not located in space.

In contrast with epigenesis and tautology, which constitute the worlds of replication, there is the whole realm of creativity, art, learning, and evolution, in which the ongoing processes of change *feed on the random.* The essence of epigenesis is predictable repetition; the essence of learning and evolution is exploration and change.

In the transmission of human culture, people always attempt to replicate, to pass on to the next generation the skills and values of the parents; but the attempt always and inevitably fails because cultural transmission is geared to learning, not to DNA. The process of transmission of culture is a sort of hybrid or mix-up of the two realms. It must attempt to use the phenomena of learning for the purpose of replication because what the parents have was learned by them. If the offspring miraculously had the DNA that would give them the parental skills, those skills would be *different* and perhaps nonviable.

It is interesting that between the two worlds is the cultural phenomenon of *explanation*—the mapping onto* tautology of unfamiliar sequences of events.

Finally, it will be noted that the realms of epigenesis and of evolution are, at a deeper level, typified in the twin paradigms of the second law of thermodynamics: (1) that the random workings of probability will always eat up order, pattern, and negative entropy but (2) that for the creation of new order, the workings of the random, the plethora of uncommitted alternatives (entropy) is necessary. It is out of the random that organisms collect new mutations, and it is there that stochastic learning gathers its solutions. Evolution leads to climax: ecological saturation of all the possibilities of differentiation. Learning leads to the overpacked mind. By return to the unlearned and mass-produced egg, the ongoing species again and again clears its memory banks to be ready for the new.

---

* I use the phrase, *to map onto,* for the following reasons: All description, explanation, or representation is necessarily in some sense a mapping of derivatives from the phenomena to be described onto some surface or matrix or system of coordinates. In the case of an actual map, the receiving matrix is commonly a flat sheet of paper of finite extent, and difficulties occur when that which is to be mapped is too big or, for example, spherical. Other difficulties would be generated if the receiving matrix were the surface of a torus (doughnut) or if it were a discontinuous lineal sequence of points. Every receiving matrix, even a language or a tautological network of propositions, will have its formal characteristics which will *in principle* be distortive of the phenomena to be mapped onto it. The universe was, perhaps, designed by Procrustes, that sinister character of Greek mythology in whose inn every traveler had to fit the bed on pain of amputation or elongation of the legs.

9. NUMBER IS DIFFERENT FROM QUANTITY

This difference is basic for any sort of theorizing in behavioral science, for any sort of imagining of what goes on between organisms or inside organisms as part of their processes of thought.

Numbers are the product of counting. *Quantities* are the product of measurement. This means that numbers can conceivably be accurate because there is a discontinuity between each integer and the next. Between *two* and *three,* there is a jump. In the case of quantity, there is no such jump; and because jump is missing in the world of quantity, it is impossible for any quantity to be exact. You can have exactly three tomatoes. You can never have exactly three gallons of water. Always quantity is approximate.

Even when number and quantity are clearly discriminated, there is another concept that must be recognized and distinguished from both number and quantity. For this other concept, there is, I think, no English word, so we have to be content with remembering that there is a subset of *patterns* whose members are commonly called "numbers." Not all numbers are the products of counting. Indeed, it is the smaller, and therefore commoner, numbers that are often not counted but recognized as patterns at a single glance. Cardplayers do not stop to count the pips in the eight of spades and can even recognize the characteristic patterning of pips up to "ten."

In other words, number is of the world of pattern, gestalt, and digital computation; quantity is of the world of analogic and probabilistic computation.

Some birds can somehow distinguish number up to seven. But whether this is done by counting or by pattern recognition is not known. The experiment that came closest to testing this difference between the two methods was performed by Otto Koehler with a jackdaw. The bird was trained to the following routine: A number of small cups with lids are set out. In these cups, small pieces of meat are placed. Some cups have one piece of meat, some have two or three, and some cups have none. Separate from the cups, there is a plate on which there is a number of pieces of meat greater than the total number of pieces in the cups. The jackdaw learns to open each cup, taking off the lid, and then eats any pieces of meat that are in the cup. Finally, when he has eaten all the meat in the cups, he may go to the plate and there eat the *same number* of pieces of meat that he got from the cups. The bird is punished if he eats more meat from the plate than was in the cups. This routine he is able to learn.

Now, the question is: Is the jackdaw counting the pieces of meat, or is he using some alternative method of identifying the number of pieces? The experiment has been carefully designed to push the bird toward counting. His actions are interrupted by his having to lift the lids, and the sequence has been further confused by having some cups contain more than one piece of meat and some contain none. By these devices, the experimenter has tried to make it impossible for the jackdaw to create some sort of pattern or rhythm by which to recognize the number of the pieces of meat. The bird is thus forced, so far as the experimenter could force the matter, to count the pieces of meat.

It is still conceivable, of course, that the taking of the meat from the cups becomes some sort of rhythmic dance and that this rhythm is in some way repeated when the bird takes the meat from the plate. The matter is still conceivably in doubt, but on the whole, the experiment is rather convincing in favor of the hypothesis that the jackdaw is counting the pieces of meat rather than recognizing a pattern either of pieces or of his own actions.

It is interesting to look at the biological world in terms of this question: Should the various instances in which number is exhibited be regarded as instances of gestalt, of counted number, or of mere quantity? There is a rather conspicuous difference between, for example, the statement "This single rose has five petals, and it has five sepals, and indeed its symmetry is of a pentad pattern" and the statement "This rose has one hundred and twelve stamens, and that other has ninety-seven, and this has only sixty-four." The process which controls the number of stamens is surely different from the process that controls the number of petals or sepals. And, interestingly, in the double rose, what seems to have happened is that some of the stamens have been converted into petals, so that the process for determining how many petals to make has now become, not the normal process delimiting petals to a pattern of five, but more like the process determining the *quantity* of stamens. We may say that petals are normally "five" in the single rose but that stamens are "many" where "many" is a quantity that will vary from one rose to another.

With this difference in mind, we can look at the biological world and ask what is the largest number that the processes of growth can handle as a fixed pattern, beyond which the matter is handled as quantity. So far as I know, the "numbers" two, three, four, and five are the common ones in the symmetry of plants and animals, particularly in radial symmetry.

The reader may find pleasure in collecting cases of rigidly controlled or patterned numbers in nature. For some reason, the larger numbers seem to be confined to linear series of segments, such as the vertebrae of mammals, the abdominal segments of insects, and the anterior segmentation of earthworms. (At the front end, the segmentation is rather rigidly controlled down to the segments bearing genital organs. The numbers vary with the species but may reach fifteen. After that, the tail has "many" segments.) An interesting addition to these observations is the common circumstance that an organism, having chosen a number for the radial symmetry of some set of parts, will repeat that number in other parts. A lily has three sepals and then three petals and then six stamens and a trilocular ovary.

It appears that what seemed to be a quirk or peculiarity of human operation—namely, that we occidental humans get numbers by counting or pattern recognition while we get quantities by measurement—turns out to be some sort of universal truth. Not only the jackdaw but also the rose are constrained to show that for them, too—for the rose in its anatomy and for the jackdaw in its behavior (and, of course, in its vertebral segmentation)—there is this profound difference between numbers and quantity.

What does this mean? That question is very ancient and certainly goes back to Pythagoras, who is said to have encountered a similar regularity in the relation between harmonics.

The hexago-rectangle discussed in section 5 provides a means of posing these questions. We saw, in that case, that the components of description could be quite various. In that particular case, to attach more validity to one rather than to another *way of organizing* the description would be to indulge illusion. But in this matter of biological numbers and quantities, it seems that we encounter something more profound. Does this case differ from that of the hexago-rectangle? And if so, how?

I suggest that neither case is as trivial as the problems of the hexago-rectangle seemed to be at first sight. We go back to the eternal verities of Saint Augustine: "Listen to the thunder of that saint, in about A.D. 500: 7 and 3 are 10; 7 and 3 have always been 10; 7 and 3 at no time and in no way have ever been anything but 10; 7 and 3 will always be 10." *

No doubt, in asserting the contrast between numbers and quantities, I am close to asserting an eternal verity, and Augustine would surely agree.

But we can reply to the saint, "Yes, very true. But is that really what you want and mean to say? It is also true, surely, that 3 and 7 are 10, and that 2 and 1 and 7 are 10, and that 1 and 1 and 1 and 1 and 1 and 1 and 1 and 1 and 1 and 1 are 10. In fact, the eternal verity that you are trying to assert is much more general and profound than the special case used by you to carry that profound message." But we can agree that the more abstract eternal verity will be difficult to state with unambiguous precision.

In other words, it is possible that many of the ways of describing my hexago-rectangle could be only different surfacings of the same more profound and more general tautology (where Euclidean geometry is viewed as a tautological system).

It is, I think, correct to say, not only that the various phrasings of the description of the hexago-rectangle ultimately agree about what the describers thought they saw but also that there is an agreement about a single more general and profound tautology in terms of which the various descriptions are organized.

In this sense, the distinction between numbers and quantities is, I believe, nontrivial and is shown to be so by the anatomy of the rose with its "5" petals and its "many" stamens, and I have put quotation marks into my description of the rose to suggest that the names of the numbers and of the quantities are the surfacing of formal ideas, immanent within the growing rose.

* So quoted by Warren McCulloch in *Embodiments of Mind* (Cambridge: MIT Press, 1965).

10. QUANTITY DOES NOT DETERMINE PATTERN

It is impossible, in principle, to explain any pattern by invoking a single quantity. But note that *a ratio between two quantities* is already the beginning of pattern. In other words, quantity and pattern are of different logical type* and do not readily fit together in the same thinking.

What appears to be a genesis of pattern by quantity arises where the pattern was latent before the quantity had impact on the system. The familiar case is that of tension which will break a chain at the weakest link. Under change of a quantity, tension, a latent difference is made manifest or, as the photographers would say, developed. The development of a photographic negative is precisely the making manifest of latent differences laid down in the photographic emulsion by previous differential exposure to light.

Imagine an island with two mountains on it. A quantitative change, a rise, in the level of the ocean may convert this single island into two islands. This will happen at the point where the level of the ocean rises higher than the saddle between the two mountains. Again, the qualitative pattern was latent before the quantity had impact on it; and when the pattern changed, the change was sudden and discontinuous.

There is a strong tendency in explanatory prose to invoke quantities of tension, energy, and whatnot to explain the genesis of pattern. I believe that all such explanations are inappropriate or wrong. From the point of view of any agent who imposes a quantitative change, any change of pattern which may occur will be unpredictable or divergent.

11. THERE ARE NO MONOTONE "VALUES" IN BIOLOGY

A monotone value is one that either only increases or only decreases. Its curve has no kinks; that is, its curve never changes from

---

* Bertrand Russell's concept of logical type will be discussed in some detail later, especially in the final section of Chapter 4. For the present, understand that because a *class* cannot be a member of itself, conclusions that can be drawn only from multiple cases (e.g., from differences between pairs of items) are of different logical type from conclusions drawn from a single item (e.g., from a quantity). (Also see Glossary.)

increase to decrease or vice versa. Desired substances, things, patterns, or sequences of experience that are in some sense "good" for the organism—items of diet, conditions of life, temperature, entertainment, sex, and so forth—are never such that more of the something is always better than less of the something. Rather, for all objects and experiences, there is a quantity that has optimum value. Above that quantity, the variable becomes toxic. To fall below that value is to be deprived.

This characteristic of biological value does not hold for money. Money is always transitively valued. More money is supposedly always better than less money. For example, $1,001 is to be preferred to $1,000. But this is not so for biological values. More calcium is not always better than less calcium. There is an optimum quantity of calcium that a given organism may need in its diet. Beyond this, calcium becomes toxic. Similarly, for oxygen that we breathe or foods or components of diet and probably all components of relationship, enough is better than a feast. We can even have too much psychotherapy. A relationship with no combat in it is dull, and a relationship with too much combat in it is toxic. What is desirable is a relationship with a certain optimum of conflict. It is even possible that when we consider money, not by itself, but as acting on human beings who own it, we may find that money, too, becomes toxic beyond a certain point. In any case, the philosophy of money, the set of presuppositions by which money is supposedly better and better the more you have of it, is totally antibiological. It seems, nevertheless, that this philosophy can be taught to living things.

12. SOMETIMES SMALL IS BEAUTIFUL

Perhaps no variable brings the problems of being alive so vividly and clearly before the analyst's eye as does size. The elephant is afflicted with the problems of bigness; the shrew, with those of smallness. But for each, there is an optimum size. The elephant would not be better off if he were much smaller, nor would the shrew be relieved by being much bigger. We may say that each is *addicted* to the size that is.

There are purely physical problems of bigness or smallness, problems that affect the solar system, the bridge, and the wristwatch. But in addition to these, there are problems special to aggregates of living matter, whether these be single creatures or whole cities.

Let us first look at the physical. Problems of mechanical *instability* arise because, for example, the forces of gravity do not follow the same quantitative regularities as those of cohesion. A large clod of earth is easier to break by dropping it on the ground than is a small one. The glacier grows and therefore, partly melting and partly breaking, must begin a changed existence in the form of avalanches, smaller units that must fall off the larger matrix. Conversely, even in the physical universe, the very small may become unstable *because* the relation between surface area and weight is nonlinear. We break up any material which we wish to dissolve because the smaller pieces have a greater ratio of surface to volume and will therefore give more access to the solvent. The larger lumps will be the last to disappear. And so on.

To carry these thoughts over into the more complex world of living things, a fable may be offered:

THE TALE OF THE POLYPLOID HORSE

They say the Nobel people are still embarrassed when anybody mentions polyploid horses. Anyhow, Dr. P. U. Posif, the great Erewhonian geneticist, got his prize in the late 1980s for jiggling with the DNA of the common cart horse (*Equus caballus*). It was said that he made a great contribution to the then new science of transportology. At any rate, he got his prize for *creating*—no other word would be good enough for a piece of applied science so nearly usurping the role of deity—creating, I say, a horse precisely twice the size of the ordinary Clydesdale. It was twice as long, twice as high, and twice as thick. It was a polyploid, with four times the usual number of chromosomes.

P. U. Posif always claimed that there was a time, when this wonderful animal was still a colt, when it was able to stand on its four legs. A wonderful sight it must have been! But anyhow, by the time the horse was shown to the public and recorded with all the communicational devices of modern civilization, the horse was not doing any standing. In a word, it was *too heavy*. It weighed, of course, eight times as much as a normal Clydesdale.

For a public showing and for the media, Dr. Posif always insisted on turning off the hoses that were continuously necessary to keep the beast at normal mammalian temperature. But we were always afraid that the innermost parts would begin to cook. After all, the poor beast's skin and dermal fat were twice as thick as normal, and its surface area was only four times that of a normal horse, so it didn't cool properly.

Every morning, the horse had to be raised to its feet with the aid of a small crane and hung in a sort of box on wheels, in which it was suspended on springs, adjusted to take half its weight off its legs.

Dr. Posif used to claim that the animal was outstandingly intelligent. It had, of course, eight times as much brain (by weight) as any other horse, but I could never see that it was concerned with any questions more complex than those which interest other horses. It had very little free time, what with one thing and another—always panting, partly to keep cool and partly to oxygenate its eight-times body. Its windpipe, after all, had only four times the normal area of cross section.

And then there was eating. Somehow it had to eat, every day, eight times the amount that would satisfy a normal horse and had to push all that food down an esophagus only four times the caliber of the normal. The blood vessels, too, were reduced in relative size, and this made circulation more difficult and put extra strain on the heart.

A sad beast.

The fable shows what inevitably happens when two or more variables, whose curves are discrepant, interact. That is what produces the interaction between change and tolerance. For instance, gradual growth in a population, whether of automobiles or of people, has no perceptible effect upon a transportation system until *suddenly* the threshold of tolerance is passed and the traffic jams. The changing of one variable exposes a critical value of the other.

Of all such cases, the best known today is the behavior of fissionable material in the atom bomb. The uranium occurs in nature and is continually undergoing fission, but no explosion occurs because no chain reaction is established. Each atom, as it breaks, gives off neutrons that, if they hit another uranium atom, may cause fission, but many neutrons are merely lost. Unless the lump of uranium is of critical size, an average of less than one neutron from each fission will break another atom, and the chain will dwindle. If the lump is made bigger, a larger fraction of the neutrons will hit uranium atoms to cause fission. The process will then achieve positive exponential *gain* and become an explosion.

In the case of the imaginary horse, length, surface area, and volume (or mass) become discrepant because their curves of increase have mutually nonlinear characteristics. Surface varies as the square of length, volume varies as the cube of length, and surface varies as the ⅔ power of volume.

For the horse (and for all real creatures), the matter becomes more serious because to remain alive, many internal motions must be maintained. There is an internal logistics of blood, food, oxygen, and excretory products and a logistics of information in the form of neural and hormonal messages.

The harbor porpoise, which is about three feet long, with a jacket of blubber about one inch thick and a surface area of about six square feet, has a known heat budget that balances comfortably in Arctic waters. The heat budget of a big whale, which is about ten times the length of the porpoise (i.e., 1,000 times the volume and 100 times the surface), with a blubber jacket nearly twelve inches thick, is totally mysterious. Presumably, they have a superior logistic system moving blood through the dorsal fins and tail flukes, where all cetaceans get rid of heat.

The fact of growth adds another order of complexity to the problems of bigness in living things. Will growth alter the proportions of the organism? These problems of the limitation of growth are met in very different ways by different creatures.

A simple case is that of the palms, which do not adjust their girth to compensate for their height. An oak tree with growing tissue (cambium) between its wood, and its bark grows in length and width throughout its life. But a coconut palm, whose only growing tissue is at the apex of the trunk (the so-called millionaire's salad, which can only be got at the price of killing the palm), simply gets taller and taller, with some slow increase of the bole at its base. For this organism, the limitation of height is simply a normal part of its adaptation to a niche. The sheer mechanical instability of excessive height without compensation in girth provides its normal way of death.

Many plants avoid (or solve?) these problems of the limitation of growth by linking their life-span to the calendar or to their own reproductive cycle. Annuals start a new generation each year, and plants like the so-called century plant (yucca) may live many years but, like the salmon, inevitably die when they reproduce. Except for multiple branching within the flowering head, the yucca makes no branches. The branching inflorescence itself is its terminal stem; when that has completed its function, the plant dies. Its death is normal to its way of life.

Among some higher animals, growth is controlled. The creature reaches a size or age or stage at which growth simply stops (i.e., is stopped by chemical or other messages within the organization of the creature). The cells, under control, cease to grow and divide. When controls no longer operate (by failure to generate the message or failure to receive it), the result is cancer. Where do such messages originate, what triggers their sending, and in what presumably chemical code are these messages immanent? What controls the nearly perfect external bilateral symmetry of the mammalian body? We have remarkably little knowledge of the message system that controls growth. There must be a whole interlocking system as yet scarcely studied.

13. LOGIC IS A POOR MODEL OF CAUSE AND EFFECT

We use the same words to talk about logical sequences and about sequences of cause and effect. We say, *"If* Euclid's definitions and postulates are accepted, *then* two triangles having three sides of the one equal to three sides of the other are equal each to each." And we say, *"If* the temperature falls below 0°C, *then* the water begins to become ice."

But the *if . . . then* of logic in the syllogism is very different from the *if . . . then* of cause and effect.

In a computer, which works by cause and effect, with one transistor triggering another, the sequences of cause and effect are used to *simulate* logic. Thirty years ago, we used to ask: Can a computer simulate *all* the processes of logic? The answer was yes, but the question was surely wrong. We should have asked: Can logic simulate all sequences of cause and effect? And the answer would have been no.

When the sequences of cause and effect become circular (or more complex than circular), then the description or mapping of those sequences onto timeless logic becomes self-contradictory. Paradoxes are generated that pure logic cannot tolerate. An ordinary buzzer circuit will serve as an example, a single instance of the apparent paradoxes gen-

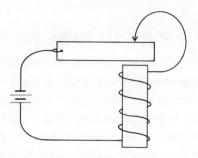

Figure 3

erated in a million cases of homeostasis throughout biology. The buzzer circuit (see Figure 3) is so rigged that current will pass around the circuit when the armature makes contact with the electrode at *A*. But the passage of current activates the electromagnet that will draw the armature away, breaking the contact at *A*. The current will then cease to pass around the circuit, the electromagnet will become inactive, and the armature will return to make contact at *A* and so repeat the cycle.

If we spell out this cycle onto a causal sequence, we get the following:

> If contact is made at *A*, then the magnet is activated.
> If the magnet is activated, then contact at *A* is broken.
> If contact at *A* is broken, then the magnet is inactivated.
> If magnet is inactivated, then contact is made.

This sequence is perfectly satisfactory provided it is clearly understood that the *if . . . then* junctures are *causal*. But the bad pun that would move the *if*s and *then*s over into the world of logic will create havoc:

> If the contact is made, then the contact is broken.
> If *P*, then not *P*.

The *if . . . then* of causality contains *time*, but the *if . . . then* of logic is timeless. It follows that logic is an incomplete model of causality.

14. CAUSALITY DOES NOT WORK BACKWARD

Logic can often be reversed, but the effect does not precede the cause. This generalization has been a stumbling block for the psychological and biological sciences since the times of Plato and Aristotle. The Greeks were inclined to believe in what were later called *final* causes. They believed that the pattern generated at the end of a sequence of events could be regarded as in some way causal of the pathway followed by that sequence. This led to the whole of teleology, as it was called (*telos* meaning the end or purpose of a sequence).

The problem which confronted biological thinkers was the problem of adaptation. It appeared that a crab had claws in order to hold things. The difficulty was always in arguing backward from the purpose of claws to the causation of the development of claws. For a long time, it was considered heretical in biology to believe that claws were there *because* they were useful. This belief contained the teleological fallacy, an inversion of causality in time.

Lineal thinking will always generate either the teleological fallacy (that end determines process) or the myth of some supernatural controlling agency.

What is the case is that when causal systems become circular (a matter to be discussed in Chapter 4), a change in any part of the circle can be regarded as *cause* for change at a later time in any variable anywhere in the circle. It thus appears that a rise in the temperature of the room can be regarded as the cause of the change in the switch of the thermostat and, alternatively, that the action of the thermostat can be regarded as controlling the temperature of the room.

15. LANGUAGE COMMONLY STRESSES ONLY ONE SIDE OF ANY INTERACTION

We commonly speak as though a single "thing" could "have" some characteristic. A stone, we say, is "hard," "small," "heavy," "yellow," "dense," "fragile," "hot," "moving," "stationary," "visible," "edible," "inedible," and so on.

That is how our language is made: "The stone is hard." And so on. And that way of talking is good enough for the marketplace: "That is a new brand." "The potatoes are rotten." "The eggs are fresh." "The container is damaged." "The diamond is flawed." "A pound of apples is enough." And so on.

But this way of talking is not good enough in science or epistemology. To think straight, it is advisable to expect all qualities and attributes, adjectives, and so on to refer to at least *two* sets of interactions in time.

"The stone is hard" means a) that when poked it resisted penetration and b) that certain continual interactions among the molecular *parts* of the stone in some way bond the parts together.

"The stone is stationary" comments on the location of the stone relative to the location of the speaker and other possible moving things. It also comments on matters internal to the stone: its inertia, lack of internal distortion, lack of friction at the surface, and so on.

Language continually asserts by the syntax of subject and predicate that "things" somehow "have" qualities and attributes. A more precise way of talking would insist that the "things" are produced, are seen as separate from other "things," and are made "real" by their internal relations and by their behavior in relationship with other things and with the speaker.

It is necessary to be quite clear about the universal truth that whatever "things" may be in their pleromatic and thingish world, they can only enter the world of communication and meaning by their names, their qualities and their attributes (i.e., by reports of their internal and external relations and interactions).

16. *"STABILITY"* and *"CHANGE"* DESCRIBE PARTS OF OUR DESCRIPTIONS

In other parts of this book, the word *stable* and also, necessarily, the word *change* will become very important. It is therefore wise to examine these words now in the introductory phase of our task. What traps do these words contain or conceal?

Stable is commonly used as an adjective applied to a thing. A chemical compound, house, ecosystem, or government is described as stable. If we pursue this matter further, we shall be told that the stable object is unchanging under the impact or stress of some particular external or internal variable or, perhaps, that it resists the passage of time.

If we start to investigate what lies behind this use of *stability,* we shall find a wide range of mechanisms. At the simplest level, we have simple physical hardness or viscosity, qualities descriptive of relations of impact between the stable object and some other. At more complex levels, the whole mass of interlocking processes called *life* may be involved in keeping our object in a *state of change* that can maintain some necessary constants, such as body temperature, blood circulation, blood sugar, or even life itself.

The acrobat on the high wire maintains his stability by continual correction of his imbalance.

These more complex examples suggest that when we use *stability* in talking about living things or self-corrective circuits, we should *follow the example of the entities about which we are talking.* For the acrobat on the high wire, his or her so-called "balance" is important; so, for the mammalian body, is its "temperature." The changing state of these important variables from moment to moment is reported in the communication networks of the body. To follow the example of the entity, we should define "stability" always by reference to the *ongoing truth of some descriptive proposition.* The statement "The acrobat is on the high wire" continues to be true under impact of small breezes and vibrations of the wire. This "stability" is the result of continual changes in descriptions of the acrobat's posture and the position of his or her balancing pole.

It follows that when we talk about living entities, statements about "stability" should always be labeled by reference to some descriptive proposition so that the typing of the word, *stable,* may be clear. We shall see later, especially in Chapter 4, that *every* descriptive proposition is to be characterized according to logical typing of subject, predicate, and context.

Similarly, all statements about change require the same sort of precision. Such profound saws as the French *"Plus ça change, plus c'est la même chose"* owe their wiseacre wisdom to a muddling of logical types. What "changes" and what "stays the same" are both of them descriptive propositions, but of different order.

Some comment on the list of presuppositions examined in this chapter is called for. First of all, the list is in no sense complete, and there is no suggestion that such a thing as a complete list of verities or generalities could be prepared. Is it even a characteristic of the world in which we live that such a list should be finite?

In the preparation of this chapter, roughly another dozen candidates for inclusion were dropped, and a number of others were removed from this chapter to become integrated parts of Chapters 3, 4, and 5. However, even with its incompleteness, there are a number of possible exercises that the reader might perform with the list.

First, when we have a list, the natural impulse of the scientist is to start classifying or ordering its members. This I have partly done, breaking the list into four groups in which the members are linked together in various ways. It would be a nontrivial exercise to list the ways in which such verities or presuppositions may be connected. The grouping I have imposed is as follows:

A first cluster includes numbers 1 to 5, which seem to be related aspects of the necessary phenomenon of coding. Here, for example, the proposition that "science never proves anything" is rather easily recognized as a synonym for the distinction between map and territory; both follow from the Ames experiments and the generalization of natural history that "there is no objective experience."

It is interesting to note that on the abstract and philosophical side, this group of generalizations has to depend very closely on something like Occam's razor or the rule of parsimony. Without some such ultimate criterion, there is no ultimate way of choosing between one hypothesis and another. The criterion found necessary is of simplicity *versus* complexity. But along with these generalizations stands their connection with neurophysiology, Ames experiments, and the like. One wonders immediately whether the material on perception does not go along with the more philosophical material because the process of perception contains something like an Occam's razor or a criterion of parsimony. The discussion of wholes and parts in number 5 is a spelling out of a common form of transformation that occurs in those processes we call *description*.

Numbers 6, 7, and 8 form a second cluster, dealing with questions of the random and the ordered. The reader will observe that the notion that the new can be plucked only out of the random is in almost total contradiction to the inevitability of entropy. The whole matter of entropy and negentropy (see Glossary) and the contrasts between the set of generalities associated with these words and those associated with energy will be dealt with in Chapter 6 in the discussion of the economics of flexibility. Here it is only necessary to note the interesting formal analogy between the apparent contradiction in this cluster and the discrimination drawn in the third cluster, in which number 9 contrasts number with quantity. The sort of thinking that deals with quantity resembles in many ways the thinking that surrounds the concept of energy; whereas the concept of number is much more closely related to the concepts of pattern and negentropy.

The central mystery of evolution lies, of course, in the contrast between statements of the second law of thermodynamics and the observation that the new can only be plucked from the random. It was this contrast that Darwin partly resolved by his theory of natural selection.

The other two clusters in the list as given are 9 to 12 and 13 to 16. I will leave it to the reader to construct his or her phrasings of how these clusters are internally related and to create other clusters according to his/her own ways of thought.

In Chapter 3 I shall continue to sketch in the background of my thesis with a listing of generalities or presuppositions. I shall, however, come closer to the central problems of thought and evolution, trying to give answers to the question: *In what ways can two or more items of information or command work together or in opposition?* This question with its multiple answers seems to me to be central to any theory of thought or evolution.

III

MULTIPLE
VERSIONS
OF
THE WORLD

What I tell you three times is true.
—LEWIS CARROLL, *The Hunting of the Snark*

 Chapter 2, "Every Schoolboy Knows . . ." has introduced the reader to a number of basic ideas about the world, elementary propositions or verities with which every serious epistemology or epistemologist must make peace.

In this chapter, I go on to generalizations that are somewhat more complex in that the question which I ask takes the immediate, exoteric form: "What bonus or increment of knowing follows from *combining* information from two or more sources?"

The reader may take the present chapter and Chapter 5 "Multiple Versions of Relationship" as just two more items which the schoolboy should know. And in fact, in the writing of the book, the heading "Two descriptions are better than one" originally covered all this material. But as the more or less experimental writing of the book went on over about three years, this heading aggregated to itself a very considerable range of sections, and it became evident that the combination of diverse pieces of information defined an approach of very great power to what I call (in Chapter 1) "the pattern which connects." Particular facets of the great pattern were brought to my attention by particular ways in which two or more pieces of information could be combined.

In the present chapter, I shall focus on those varieties of combination which would seem to give the perceiving organism information about the world around itself or about itself as a part of that external world (as when the creature sees its own toe). I shall leave for Chapter 5 the more subtle and, indeed, more biological or creatural combinations that would give the perceiver more knowledge of the internal relations and processes called the *self*.

In every instance, the primary question I shall ask will concern the bonus of understanding which the combination of information affords. The reader is, however, reminded that behind the simple, superficial question there is partly concealed the deeper and perhaps mystical question, "Does the study of this particular case, in which an insight develops from the comparison of sources, throw any light on how the universe is integrated?" My method of procedure will be to ask about the immediate bonus in each case, but my ultimate goal is an inquiry into the larger pattern which connects.

1. THE CASE OF DIFFERENCE

Of all these examples, the simplest but the most profound is the fact that it takes at least two somethings to create a difference. To produce news of difference, i.e., *information,* there must be two entities (real or imagined) such that the difference between them can be immanent in their mutual relationship; and the whole affair must be such that news of their difference can be represented as a difference inside some information-processing entity, such as a brain or, perhaps, a computer.

There is a profound and unanswerable question about the nature of those "at least two" things that between them generate the difference which becomes information by making a difference. Clearly each alone is—for the mind and perception—a non-entity, a non-being. Not different from being, and not different from non-being. An unknowable, a *Ding an sich,* a sound of one hand clapping.

The stuff of sensation, then, is a pair of values of some variable, presented over a time to a sense organ whose response depends upon the ratio between the members of the pair. (This matter of the nature of difference will be discussed in detail in Chapter 4, criterion 2.)

2. THE CASE OF BINOCULAR VISION

Let us consider another simple and familiar case of double description. What is gained by comparing the data collected by one eye with the data collected by the other? Typically, both eyes are aimed at the same region of the surrounding universe, and this might seem to be a wasteful use of the sense organs. But the anatomy indicates that very considerable advantage must accrue from this usage. The innervation of the two retinas and the creation at the optic chiasma of pathways for the redistribution of information is such an extraordinary feat of morphogenesis as must surely denote great evolutionary advantage.

In brief, each retinal surface is a nearly hemispherical cup into which a lens focuses an inverted image of what is being seen. Thus, the image of what is over to the left front will be focused onto the outer side of the right retina and onto the inner side of the left retina. What is surprising is that the innervation of each retina is divided into two systems by a sharp vertical boundary. Thus, the information carried by optic fibers from the outside of the right eye meets, in the right brain, with the information carried by fibers from the inner side of the left eye. Similarly, information from the outside of the left retina and the inside of the right retina is gathered in the left brain.

The binocular image, which appears to be undivided, is in fact a complex synthesis of information from the left front in the right brain and a corresponding synthesis of material from the right front in the left brain. Later these two synthesized aggregates of information are themselves synthesized into a single subjective picture from which all traces of the vertical boundary have disappeared.

From this elaborate arrangement, two sorts of advantage accrue. The seer is able to improve resolution at edges and contrasts; and better

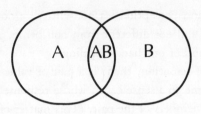

Figure 4

able to read when the print is small or the illumination poor. More important, information about depth is created. In more formal language, the *difference* between the information provided by the one retina and that provided by the other is itself information of a *different logical type*. From this new sort of information, the seer adds an extra *dimension* to seeing.

In Figure 4, let *A* represent the class or set of components of the aggregate of information obtained from some first source (e.g., the right eye), and let *B* represent the class of components of the information obtained from some second source (e.g., the left eye). Then *AB* will represent the class of components referred to by information from both eyes. *AB* must either contain members or be empty.

If there exist real members of *AB,* then the information from the second source has imposed a subclassification upon *A* that was previously impossible (i.e., has provided, in combination with *A,* a logical type of information of which the first source alone was incapable).

We now proceed with the search for other cases under this general rubric and shall specifically look in each case for the genesis of information of new logical type out of the juxtaposing of multiple descriptions. In principle, extra "depth" in some metaphoric sense is to be expected whenever the information for the two descriptions is differently collected or differently coded.

3. THE CASE OF THE PLANET PLUTO

Human sense organs can receive *only* news of difference, and the differences must be coded into events in *time* (i.e., into *changes*) in order to be perceptible. Ordinary static differences that remain constant for more than a few seconds become perceptible only by scanning. Similarly, very slow changes become perceptible only by a combination of scanning *and* bringing together observations from separated moments in the continuum of time.

An elegant (i.e., an economical) example of these principles is provided by the device used by Clyde William Tombaugh, who in 1930, while still a graduate student, discovered the planet Pluto.

From calculations based on disturbances in the orbit of Neptune it seemed that these irregularities could be explained by gravitational pull from some planet in an orbit outside the orbit of Neptune. The calculations indicated in what region of the sky the new planet could be expected at a given time.

The object to be looked for would certainly be very small and dim (about 15th magnitude), and its appearance would differ from that of other objects in the sky only in the fact of very slow movement, so slow as to be quite imperceptible to the human eye.

This problem was solved by the use of an instrument which astronomers call a *blinker*. Photographs of the appropriate region of the sky were taken at longish intervals. These photographs were then studied in pairs in the blinker. This instrument is the converse of a binocular microscope; instead of two eyepieces and one stage, it has one eyepiece and two stages and is so arranged that by the flick of a lever, what is seen at one moment on one stage can be replaced by a view of the other stage. Two photographs are placed in exact register on the two stages so that all the ordinary fixed stars precisely coincide. Then, when the lever is flicked over, the fixed stars will not appear to move, but a planet will appear to jump from one position to another. There were, however, many jumping objects (asteroids) in the field of the photographs, and Tombaugh had to find one that jumped *less* than the others.

After hundreds of such comparisons, Tombaugh saw Pluto jump.

4. THE CASE OF SYNAPTIC SUMMATION

Synaptic summation is the technical term used in neurophysiology for those instances in which some neuron C is fired only by a combination of neurons A and B. A alone is insufficient to fire C, and B alone is insufficient to fire C; but if neurons A and B fire together within a lim-

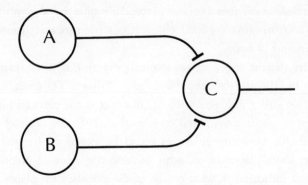

Figure 5

ited period of microseconds, then C is triggered (see Figure 5). Notice that the conventional term for this phenomenon, *summation,* would suggest an *adding* of information from one source to information from another. What actually happens is not an adding but a forming of a logical product, a process more closely akin to multiplication.

What this arrangement does to the information that neuron A alone could give is a segmentation or subclassification of the firings of A into two classes, namely, those firings of A accompanied by B and those firings of A which are not accompanied by B. Correspondingly, the firings of neuron B are subdivided into two classes, those accompanied by A and those not accompanied by A.

5. THE CASE OF THE HALLUCINATED DAGGER

Macbeth is about to murder Duncan, and in horror at his deed, he hallucinates a dagger (Act II, scene I).

> *Is this a dagger which I see before me,*
> *The handle toward my hand? Come, let me clutch thee.*
> *I have thee not, and yet I see thee still.*
> *Art thou not, fatal vision, sensible*
> *To feeling as to sight? or art thou but*
> *A dagger of the mind, a false creation,*
> *Proceeding from the heat-oppressed brain?*
> *I see thee yet, in form as palpable*

As this which now I draw.
Thou marshall'st me the way that I was going;
And such an instrument I was to use.
Mine eyes are made the fools o' the other senses,
Or else worth all the rest: I see thee still;
And on thy blade and dudgeon gouts of blood,
Which was not so before. There's no such thing:
It is the bloody business which informs
Thus to mine eyes.

This literary example will serve for all those cases of double description in which data from two or more different senses are combined. Macbeth "proves" that the dagger is only an hallucination by checking with his sense of touch, but even that is not enough. Perhaps his eyes are "worth all the rest." It is only when "gouts of blood" appear on the hallucinated dagger that he can dismiss the whole matter: "There's no such thing."

Comparison of information from one sense with information from another, combined with change in the hallucination, has offered Macbeth the metainformation that his experience was imaginary. In terms of Figure 4, *AB* was an empty set.

6. THE CASE OF SYNONYMOUS LANGUAGES

In many cases, an increment of insight is provided by a second language of description without the addition of any extra so-called objective information. Two proofs of a given mathematical theorem may combine to give the student an extra grasp of the relation which is being demonstrated.

Every schoolboy knows that $(a + b)^2 = a^2 + 2ab + b^2$, and he may be aware that this algebraic equation is a first step in a massive branch of mathematics called *binomial theory*. The equation itself is sufficiently demonstrated by the algorithm of algebraic multiplication, each step of which is in accord with the definitions and postulates of the tautology called *algebra*—that tautology whose subject matter is the expansion and analysis of the notion "any."

But many schoolboys do not know that there is a geometric

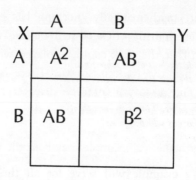

Figure 6

demonstration of the same binomial expansion (see Figure 6). Consider the straight line XY, and let this line be composed of two segments, a and b. The line is now a geometric representation of $(a+b)$ and the square constructed upon XY will be $(a+b)^2$; that is, it will have an *area called "$(a+b)^2$."*

This square can now be dissected by marking off the length a along the line XY and along one of the adjacent sides of the square and completing the figure by drawing the appropriate lines parallel to the sides of the square. The schoolboy can now think that he sees that the square is cut up into four pieces. There are two squares, one of which is a^2 while the other is b^2, and two rectangles, each of which is of area $(a \times b)$ (i.e., $2ab$).

Thus, the familiar algebraic equation $(a+b)^2 = a^2 + 2ab + b^2$ also seems to be true in Euclidean geometry. But surely it was too much to hope for that the separate pieces of the quantity $a^2 + 2ab + b^2$ would still be neatly separate in the geometric translation.

But what has been said? By what right did we substitute a so-called "length" for a and another for b and assume that, placed end to end, they would make a straight line $(a+b)$ and so on? Are we *sure* that the lengths of lines obey arithmetic rules? What has the schoolboy learned from our stating the same old equation in a new language?

In a certain sense, *nothing* has been added. No new information has been generated or captured by my asserting that $(a+b)^2 = a^2 + 2ab + b^2$ in geometry as well as in algebra.

Does a *language,* then, as such, contain *no* information?

But even if, mathematically, nothing has been added by the little mathematical conjuring trick, I still believe that the schoolboy who has never seen that the trick could be played will have a chance to learn something when the trick is shown. There is a contribution to didactic method. The discovery (if it be discovery) that the two languages (of algebra and of geometry) are mutually translatable is itself an *enlightenment*.

Another mathematical example may help the reader to assimilate the effect of using two languages.*

Ask your friends, "What is the sum of the first ten odd numbers?"

The answers will probably be statements of ignorance or attempts to add up the series:

$1 + 3 + 5 + 7 + 9 + 11 + 13 + 15 + 17 + 19$.

Show them that:

The sum of the first odd number is 1.

The sum of the first two odd numbers is 4.

The sum of the first three odd numbers is 9.

The sum of the first four odd numbers is 16.

The sum of the first five odd numbers is 25.

And so on.

Rather soon, your friends will say something like, "Oh, then the sum of the first ten odd numbers must be 100." They have learned the *trick* for adding series of odd numbers.

But ask for an explanation of why this trick *must* work and the average nonmathematician will be unable to answer. (And the state of elementary education is such that many will have no idea of how to proceed in order to create an answer.)

What has to be discovered is the difference between the *ordinal name* of the given odd number and its *cardinal* value—a difference in logical type! We are accustomed to expect that the name of a numeral

---

* I am indebted to Gertrude Hendrix for this, to most people, unfamiliar regularity: Gertrude Hendrix, "Learning by Discovery," *The Mathematics Teacher* 54 (May 1961): 290–299.

will be the same as its numerical value.* But indeed, in this case, the name is not the same as the thing named.

The sum of the first three odd numbers is 9. That is, the sum is the *square of the ordinal name* (and in this case, the ordinal name of 5 is "3") of the largest number in the series to be summed. Or—if you like—it is the square of the *number of numbers* in the series to be summed. This is the verbal statement of the trick.

To prove that the trick will work, we have to show that the difference between two consecutive summations of odd numbers is equal and *always* equal to the difference between the squares of their ordinal names.

For example, the sum of the first five odd numbers *minus* the sum of first four odd numbers must equal $5^2 - 4^2$. At the same time, we must notice that, of course, the difference between the two sums is indeed the odd number that was last added to the stack. In other words, this last added number must be equal to the difference between the squares.

Consider the same matter in a visual language. We have to demonstrate that the *next* odd number will always add to the sum of the previous odd numbers just enough to make the next total equal the square of the ordinal name of that odd number.

Represent the first odd number (1) with a unit square:

1

Represent the second odd number (3) with three unit squares:

3

Add this to the previous figure:

$1+3 \approx 4$

---

* Alternatively, we may say that the number of numbers in a set is not the same as the sum of numbers in the same set. One way or the other, we encounter a discontinuity in logical typing.

Represent the third odd number (5) with five unit squares:

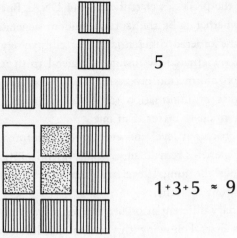

5

1₊3₊5 ≈ 9

Figure 7

And so on.

That is, $4 + 5 = 9$.

And so on. The visual presentation makes it rather easy to combine ordinals, cardinals, and the regularities of summing the series.

What has happened is that the use of a system of geometric metaphor has enormously facilitated understanding of *how* the mechanical trick comes to be a rule or regularity. More important, the student has been made aware of the contrast between applying a trick and understanding the necessity of truth behind the trick. And still more important, the student has, perhaps unwittingly, had the experience of the leap from talking arithmetic to talking *about* arithmetic. Not *numbers* but *numbers of numbers*.

It was *then*, in Wallace Stevens's words,

That the grapes seemed fatter.
The fox ran out of his hole.

7. THE CASE OF THE TWO SEXES

Von Neumann once remarked, partly in jest, that for self-replication among machines, it would be a necessary condition that two machines should act in collaboration.

Fission with replication is certainly a basic requirement of life, whether it be for multiplication or for growth, and the biochemists now know broadly the processes of replication of DNA. But next comes differentiation, whether it be the (surely) random generation of variety in evolution or the ordered differentiation of embryology. Fission, seemingly, *must* be punctuated by fusion, a general truth which exemplifies the principle of information processing we are considering here: namely that two sources of information (often in contrasting modes or languages) are enormously better than one.

At the bacterial level and even among protozoa and some fungi and algae, the gametes remain superficially identical; but in all metazoa and plants above the fungal level, the *sexes* of the gametes are distinguishable one from the other.

The binary differentiation of gametes, usually one sessile and one mobile, comes first. Following this comes the differentiation into two kinds of the multicellular individuals who are the producers of the two kinds of gametes.

Finally, there are the more complex cycles called *alternation of generations* in many plants and animal parasites.

All these orders of differentiation are surely related to the informational economics of fission, fusion, and sexual dimorphism.

So, returning to the most primitive fission and fusion, we note that the first effect or contribution of fusion to the economics of genetic information is presumably some sort of *checking.*

The process of chromosomal fusion is essentially the same in all plants and animals, and wherever it occurs, the corresponding strings of DNA material are set side by side and, in a functional sense, are *compared.* If differences between the strings of material from the respective gametes are too great, fertilization (so called) cannot occur.*

In the total process of evolution, fusion, which is the central fact of sex, has the function of limiting genetic variability. Gametes that, for whatever reason, be it mutation or other, are too different from the statistical norm are likely to meet in sexual fusion with more normal gametes of opposite sex, and in this meeting, the extremes of deviation

---

* I believe that this was first argued by C. P. Martin in his *Psychology, Evolution and Sex,* 1956. Samuel Butler (in *More Notebooks of Samuel Butler,* edited by Festing Jones) makes a similar point in discussing parthenogenesis. He argues that as dreams are to thought, so parthenogenesis is to sexual reproduction. Thought is stabilized and tested against the template of external reality, but dreams run loose. Similarly, parthenogenesis can be expected to run loose; whereas zygote formation is stabilized by the mutual comparison of gametes.

will be eliminated. (Note, in passing, that this need to eliminate deviation is likely to be imperfectly met in "incestuous" mating between gametes from closely related sources.)

But although one important function of the fusion of gametes in sexual reproduction would seem to be the limitation of deviance, it is also necessary to stress the contrary function: increasing phenotypic variety. The fusion of random pairs of gametes assures that the gene pool of the participating population will be homogeneous in the sense of being well mixed. At the same time, it assures that every viable genic combination within that pool shall be created. That is, every viable gene is tested in conjunction with as many other constellations of other genes as is possible within the limits of the participating population.

As usual in the panorama of evolution, we find that the single process is Janus-like, facing in two directions. In the present case, the fusion of gametes both places a limitation on individual deviance *and* ensures the multiple recombination of genetic material.

8. THE CASE OF BEATS AND MOIRÉ PHENOMENA

Interesting phenomena occur when two or more rhythmic patterns are combined, and these phenomena illustrate very aptly the enrichment of information that occurs when one description is combined with another. In the case of rhythmic patterns, the combination of two such patterns will generate a third. Therefore, it becomes possible to investigate an unfamiliar pattern by combining it with a known second pattern and inspecting the third pattern which they together generate.

The simplest case of what I am calling the *moiré phenomenon* is the well-known production of beats when two sounds of different frequency are combined. The phenomenon is explained by mapping onto simple arithmetic, according to the rule that if one note produces a peak in every n time units and the other has a peak in every m time units, then the combination will produce a *beat* in every $m \times n$ units when the peaks coincide. The combination has obvious uses in piano tuning. Similarly, it is possible to combine two sounds of very high frequency in order to produce beats of frequency low enough to be heard by the human ear. Sonar devices that operate on this principle are now available for the blind. A beam of high-frequency sound is emitted, and the echoes that this beam generates are received back into an "ear" in which a lower but still inaudible frequency is being generated. The resulting beats are then passed on to the human ear.

The matter becomes more complex when the rhythmic patterns, instead of being limited, as frequency is, to the single dimension of time, exist in two or more dimensions. In such cases, the result of combining the two patterns may be surprising.

Three principles are illustrated by these moiré phenomena: First, any two patterns may, if appropriately combined, generate a third. Second, any two of these three patterns could serve as base for a description of the third. Third, the whole problem of defining what is meant by the word *pattern* can be approached through these phenomena. Do we, in fact, carry around with us (like the blind person's sonar) samples of various sorts of regularity against which we can try the information (news of regular differences) that comes in from outside? Do we, for example, use our habits of what is called "dependency" to test the characteristics of other persons?

Do animals (and even plants) have characteristics such that in a given niche there is a testing of that niche by something like the moiré phenomenon?

Other questions arise regarding the nature of *aesthetic* experience. Poetry, dance, music, and other rhythmic phenomena are certainly very archaic and probably more ancient than prose. It is, moreover, characteristic of the archaic behaviors and perceptions that rhythm is continually modulated; that is, the poetry or music contains materials that could be processed by *superposing comparison* by any receiving organism with a few seconds of memory.

Is it possible that this worldwide artistic, poetical, and musical phenomenon is somehow related to moiré? If so, then the individual mind is surely deeply organized in ways which a consideration of moiré phenomena will help us to understand. In terms of the definition of "explanation" proposed in section 9, we shall say that the formal mathematics or "logic" of moiré may provide an appropriate tautology onto which these aesthetic phenomena could be mapped.

9. THE CASE OF "DESCRIPTION," "TAUTOLOGY," AND "EXPLANATION"

Among human beings, description and explanation are both highly valued, but this example of doubled information differs from most of the other cases offered in this chapter in that explanation contains no new information different from what was present in the description. Indeed, a great deal of the information that was present in description is commonly thrown away, and only a rather small part of what was to be explained is, in fact, explained. But explanation is certainly of enormous importance and certainly *seems* to give a bonus of insight over and above what was contained in description. Is the bonus of insight which explanation gives somehow related to what we got from combining two languages in section 6, above?

To examine this case, it is necessary first briefly to indicate definitions for the three words: *description, tautology,* and *explanation.*

A pure description would include all the facts (i.e., all the effective differences) immanent in the phenomena to be described but would indicate no kind of connection among these phenomena that might make them more understandable. For example, a film with sound and perhaps recordings of smell and other sense data might constitute a complete or sufficient description of what happened in front of a battery of cameras at a certain time. But that film will do little to connect the events shown on the screen one with another and will not by itself furnish any explanation. On the other hand, an explanation can be total without being descriptive. "God made everything there is" is totally explanatory but does not tell you anything about any of the things or their relations.

In science, these two types of organization of data (description and explanation) are connected by what is technically called *tautology.* Examples of tautology range from the simplest case, the assertion that "If P is true, then P is true," to such elaborate structures as the geometry of Euclid, where "If the axioms and postulates are true, then Pythagoras' theorem is true." Another example would be the axioms, definitions, postulates, and theorems of Von Neumann's Theory of Games. In such an aggregate of postulates and axioms and theorems, it is of course not claimed that any of the axioms or theorems is in any sense "true" independently or true in the outside world.

Indeed, Von Neumann, in his famous book,* expressly points out the differences between his tautological world and the more complex world of human relations. All that is claimed is that if the axioms be such and such and the postulates such and such, then the theorems will be so and so. In other words, all that the tautology affords is *connections between propositions*. The creator of the tautology stakes his reputation on the validity of these connections.

Tautology contains no information whatsoever, and explanation (the mapping of description onto tautology) contains only the information that was present in the description. The "mapping" asserts implicitly that the links which hold the tautology together correspond to relations which obtain in the description. Description, on the other hand, contains information but no logic and no explanation. For some reason, human beings enormously value this combining of ways of organizing information or material.

To illustrate how description, tautology, and explanation fit together, let me cite an assignment which I have given several times to classes. I am indebted to the astronomer Jeff Scargle for this problem, but I am responsible for the solution. The problem is:

> A man is shaving with his razor in his right hand. He looks into his mirror and in the mirror sees his image shaving with its left hand. He says, "Oh. There's been a reversal of right and left. Why is there no reversal of top and bottom?"

The problem was presented to the students in this form, and they were asked to unravel the muddle in which the man evidently is and to discuss the nature of explanation after they have accomplished this.

There are at least two twists in the problem as set. One gimmick distracts the student to focus on right and left. In fact, what has been reversed is front and back, not right and left. But there is a more subtle trouble behind that, namely, that the words *right* and *left* are not in the

* Von Neumann, J., and Morgenstern, O., *The Theory of Games and Economic Behavior* (Princeton: Princeton University Press, 1944).

same language as the words *top* and *bottom*. *Right* and *left* are words of an inner language; whereas *top* and *bottom* are parts of an external language. If the man is looking south and his image is looking north, the top is upward in himself and it is upward in his image. His east side is on the east side in the image, and his west side is on the west side in the image. *East* and *west* are in the same language as *top* and *bottom;* whereas *right* and *left* are in a different language. There is thus a logical trap in the problem as set.

It is necessary to understand that *right* and *left* cannot be defined and that you will meet with a lot of trouble if you try to define such words. If you go to the *Oxford English Dictionary*, you will find that *left* is defined as "distinctive epithet of the hand which is normally the weaker." The dictionary maker openly shows his embarrassment. If you go to Webster, you will find a more useful definition, but the author cheats. One of the rules of writing a dictionary is that you may not rely on ostensive communication for your main definition. So the problem is to define *left* without pointing to an asymmetrical object. Webster (1959) says, "that side of one's body which is toward the west when one faces north, usually the side of the less-used hand." This is using the asymmetry of the spinning earth.

In truth, the definition cannot be done without cheating. *Asymmetry* is easy to define, but there are no verbal means—and there *can* be none—for indicating which of two (mirror-image) halves is intended.

An explanation has to provide something more than a description provides and, in the end, an explanation appeals to a *tautology,* which, as I have defined it, is a body of propositions so linked together that the links *between the propositions* are necessarily valid.

The simplest tautology is "If P is true, then P is true."

A more complex tautology would be "If Q follows from P, then Q follows from P." From there, you can build up into whatever complexity you like. But you are still within the domain of the *if* clause provided, not by data, but by *you*. That is a tautology.

Now, an explanation is a mapping of the pieces of a description onto a tautology, and an explanation becomes acceptable to the degree that you are willing and able to accept the links of the tautology. If the links are "self-evident" (i.e., if they seem undoubtable to the self that is you), then the explanation built on that tautology is satisfactory to you.

That is all. It is always a matter of natural history, a matter of the faith, imagination, trust, rigidity, and so on of the organism, that is of you or me.

Let us consider what sort of tautology will serve as a foundation for our description of mirror images and their asymmetry.

Your right hand is an asymmetrical, three-dimensional object; and to define it, you require information that will link at least three polarities. To make it different from a left hand, three binary descriptive clauses must be fixed. Direction toward the palm must be distinguished from direction toward the back of the hand; direction toward the elbow must be distinguished from direction toward the fingertips; direction toward the thumb must be distinguished from direction toward the fifth finger. Now build the tautology to assert that a reversal of any one of these three binary descriptive propositions will create the mirror image (the stereo-opposite) of the hand from which we started (i.e., will create a "left" hand).

If you place your hands palm to palm so that the right palm faces north, the left will face south, and you will get a case similar to that of the man shaving.

Now, the central postulate of our tautology is that *reversal in one dimension always generates the stereo-opposite.* From this postulate, it follows—can you doubt it?—that reversal in *two* dimensions will generate the opposite of the opposite (i.e., will take us back to the form from which we started). Reversal in three dimensions will again generate the stereo-opposite. And so on.

We now flesh out our explanation by the process which the American logician, C. S. Peirce called *abduction,* that is, by finding other relevant phenomena and arguing that these, too, are cases under our rule and can be mapped onto the same tautology.

Imagine that you are an old-fashioned photographer with a black cloth over your head. You look into your camera at the ground-glass screen on which you see the face of the man whose portrait you are making. The lens is between the ground-glass screen and the subject. On the screen, you will see the image upside down and right for left but still facing you. If the subject is holding something in his right hand, he will still be holding it in his right hand on the screen but rotated 180 degrees.

If now you make a hole in the front of the camera and look in at the image formed on the ground-glass screen or on the film, the top of his head will be at the bottom. His chin will be at the top. His left will be over to the right side, *and* now he is facing himself. You have reversed three dimensions. So now you see again his stereo-opposite.

Explanation, then, consists in building a tautology, ensuring as best you can the validity of the links in the tautology so that it seems to you to be self-evident, which is in the end never totally satisfactory because nobody knows what will be discovered later.

If explanation is as I have described it, we may well wonder what bonus human beings get from achieving such a cumbersome and indeed seemingly unprofitable rigamarole. This is a question of natural history, and I believe that the problem is at least partly solved when we observe that human beings are very careless in their construction of the tautologies on which to base their explanations. In such a case, one would suppose that the bonus would be negative; but this seems not to be so, judging by the popularity of explanations which are so informal as to be misleading. A common form of empty explanation is the appeal to what I have called "dormitive principles," borrowing the word *dormitive* from Molière. There is a coda in dog Latin to Molière's *Le Malade Imaginaire,* and in this coda, we see on the stage a medieval oral doctoral examination. The examiners ask the candidate why opium puts people to sleep. The candidate triumphantly answers, "Because, learned doctors, it contains a dormitive principle."

We can imagine the candidate spending the rest of his life fractionating opium in a biochemistry lab and successively identifying in which fraction the so-called dormitive principle remained.

A better answer to the doctors' question would involve, not the opium alone, but a relationship between the opium and the people. In other words, the dormitive explanation actually falsifies the true facts of the case but what is, I believe, important is that dormitive explanations still *permit abduction.* Having enunciated a generality that opium contains a dormitive principle, it is then possible to use this type of phrasing for a very large number of other phenomena. We can say, for example, that adrenalin contains an enlivening principle and reserpine a tranquilizing principle. This will give us, albeit inaccurately and epistemologically unacceptably, handles with which to grab at a very large number of phenomena that appear to be formally comparable. And, indeed, they are formally comparable to this extent, that invoking a principle *inside one component* is in fact the error that is made in every one of these cases.

The fact remains that as a matter of natural history—and we are as interested in natural history as we are in strict epistemology—abduction is a great comfort to people, and formal explanation is often a bore. "Man thinks in two kinds of terms: one, the natural terms, shared with beasts; the other, the conventional terms (the logicals) enjoyed by man alone." *

This chapter has examined various ways in which the combining of information of different sorts or from different sources results in something more than addition. The aggregate is greater than the sum of its parts because the combining of the parts is not a simple adding but is of the nature of a multiplication or a fractionation, or the creation of a logical product. A momentary gleam of enlightenment.

So to complete this chapter and before attempting even a listing of the criteria of mental process, it is appropriate to look briefly at this structure in a much more personal and more universal way.

I have consistently held my language to an "intellectual" or "objective" mode, and this mode is convenient for many purposes (only to be avoided when used to avoid recognition of the observer's bias and stance).

To put away the quasi objective, at least in part, is not difficult, and such a change in mode is proposed by such questions as: What is this book about? What is its personal meaning to me? What am I trying to say or to discover?

The question "What am I trying to discover?" is not as unanswerable as mystics would have us believe. From the manner of the search, we can read what sort of discovery the searcher may thereby

---

* William of Ockham, 1280–1349, quoted by Warren McCulloch in his *Embodiments of Mind*, M.I.T. Press, 1965.

reach; and knowing this, we may suspect that such a discovery is what the searcher secretly and unconsciously desires.

This chapter has defined and exemplified a *manner of search,* and therefore this is the moment to raise two questions: For what am I searching? To what questions have fifty years of science led me?

The manner of the search is plain to me and might be called the *method of double or multiple comparison.*

Consider the case of binocular vision. I compared what could be seen with one eye with what could be seen with two eyes and noted that in this comparison the two-eyed method of seeing disclosed an extra dimension called *depth.* But the two-eyed way of seeing is itself an act of comparison. In other words, the chapter has been a series of comparative studies of the comparative method. The section on binocular vision (section 2) was such a comparative study of one method of comparison, and the section on catching Pluto (section 3) was another comparative study of the comparative method. Thus the whole chapter, in which such instances are placed side by side, became a display inviting the reader to achieve insight by comparing the instances one with another.

Finally, all that comparing of comparisons was built up to prepare author and reader for thought about problems of Natural Mind. There, too, we shall encounter creative comparison. It is the Platonic thesis of the book that epistemology is an indivisible, integrated metascience whose subject matter is the world of evolution, thought, adaptation, embryology, and genetics—the science of mind in the widest sense of the word.*

The comparing of these phenomena (comparing thought with evolution and epigenesis with both) is the *manner of search* of the science called "epistemology."

Or, in the phrasing of this chapter, we may say that epistemology is the bonus from combining insights from all these separate genetic sciences.

But epistemology is always and inevitably *personal.* The point of

---

*The reader will perhaps notice that consciousness is missing from this list. I prefer to use that word, not as a general term, but specifically for that strange experience whereby we (and perhaps other mammals) are sometimes conscious of the products of our perception and thought but unconscious of the greater part of the processes.

the probe is always in the heart of the explorer: What is *my* answer to the question of the nature of knowing? I surrender to the belief that my knowing is a small part of a wider integrated knowing that knits the entire biosphere or creation.

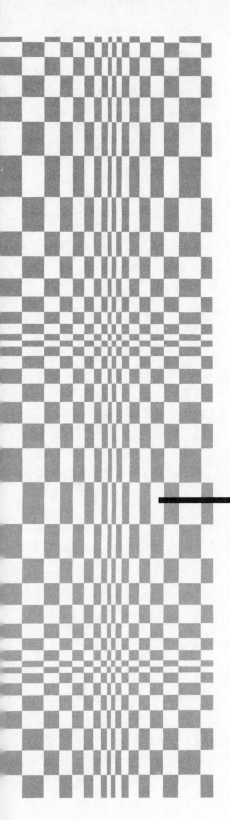

IV

CRITERIA
OF
MENTAL
PROCESS

Cogito, ergo sum.
—DESCARTES, *Discourse on Method*

 This chapter is an attempt to make a list of criteria such that if any aggregate of phenomena, any system, satisfies all the criteria listed, I shall unhesitatingly say that the aggregate is a *mind* and shall expect that, if I am to understand that aggregate, I shall need sorts of explanation different from those which would suffice to explain the characteristics of its smaller parts.

This list is the cornerstone of the whole book. No doubt other criteria could be adduced and might perhaps replace or alter the list offered here. Perhaps out of G. Spencer-Brown's *Laws of Form* or out of René Thom's *catastrophe theory,* deep restructuring of the foundations of mathematics and epistemology may come. This book must stand or fall, not by the particular content of my list, but by the validity of the idea that some such structuring of epistemology, evolution, and epigenesis is possible. I propose that the mind-body problem is soluble along lines similar to those here outlined.

The criteria of mind that seem to me to work together to supply this solution are here listed to give the reader a preliminary survey of what is proposed.

1. *A mind is an aggregate of interacting parts or components.*

2. *The interaction between parts of mind is triggered by difference,* and difference is a nonsubstantial phenomenon not located in space or time; difference is related to negentropy and entropy rather than to energy.

3. *Mental process requires collateral energy.*

4. *Mental process requires circular (or more complex) chains of determination.*

5. *In mental process, the effects of difference are to be regarded as transforms (i.e., coded versions) of events which preceded them.* The rules of such transformation must be comparatively stable (i.e., more stable than the content) but are themselves subject to transformation.

6. *The description and classification of these processes of transformation disclose a hierarchy of logical types immanent in the phenomena.*

I shall argue that the phenomena which we call *thought, evolution, ecology, life, learning,* and the like occur only in systems that satisfy these criteria.

I have already presented two considerable batches of material illustrating the nature of mental process. In Chapter 2, the reader was given almost didactic advice about how to think; and in Chapter 3, he or she was given clues to how thoughts come together. This is the beginning of a study of how to think about thinking.

We now go to use these criteria to differentiate the phenomena of *thought* from the much simpler phenomena called *material events.*

CRITERION 1. A MIND IS AN AGGREGATE OF INTERACTING PARTS OR COMPONENTS

In many cases, some parts of such an aggregate may themselves satisfy all the criteria, and in this case they, too, are to be regarded as *minds* or *subminds*. Always, however, there is a lower level of division such that the resulting parts, when considered separately, lack the complexity necessary to achieve the criteria of mind.

In a word, I do not believe that single subatomic particles are "minds" in my sense because I do believe that mental process is always a sequence of interactions *between* parts. The *explanation* of mental phenomena must always reside in the organization and interaction of multiple parts.

To many readers, it will seem unnecessary to insist upon this first criterion. But the matter is important, if only to mention and discard the contrary opinions; it is even more important to state the reasons for my intolerance. Several respected thinkers, especially Samuel Butler, to whom I have owed much pleasure and insight, and, more recently, Teilhard de Chardin, have proposed theories of evolution which assume some mental striving to be characteristic of the smallest atomies.

As I see it, these hypotheses introduce the supernatural by the back door. To accept this notion is, for me, a sort of surrender. It is saying that there are in the universe complexities of action which are inexplicable because they exist independent of any supporting complexity in which they could be supposed to be immanent. Without differentiation of parts, there can be no differentiation of events or functioning. If the atomies are not themselves internally differentiated in their individual anatomy, then the appearance of complex process can only be due to interaction between atomies.

Or if the atomies are internally differentiated, then they are by my definition *not* atomies, and I shall expect to find still simpler entities that will be devoid of mental functioning.

Finally—but only as the last resort—if de Chardin and Butler are right in supposing that the atomies have no internal differentiation and still are endowed with mental characteristics, then all explanation is impossible, and we, as scientists, should close shop and go fishing.

The whole of the present book will be based on the premise that mental function is immanent in the interaction of differentiated "parts." "Wholes" are constituted by such combined interaction.

In this matter, I prefer to follow Lamarck, who, in setting up postulates for a science of comparative psychology, laid down the rule that no mental function shall be ascribed to an organism for which the complexity of the nervous system of the organism is insufficient.*

In other words, the theory of mind presented here is holistic and, like all serious holism, is premised upon the differentiation and interaction of parts.

CRITERION 2. THE INTERACTION BETWEEN PARTS OF MIND IS TRIGGERED BY DIFFERENCE

There are, of course, many systems which are made of many parts, ranging from galaxies to sand dunes to toy locomotives. Far be it from me to suggest that all of these are minds or contain minds or engage in mental process. The toy locomotive may become a part in that mental system which includes the child who plays with it, and the galaxy may become part of the mental system which includes the astronomer and his telescope. But the objects do not become thinking subsystems in those larger minds. The criteria are useful only in combination.

We proceed now to consider the nature of the relationships between parts. How do parts interact to create mental process?

Here we meet with a very marked difference between the way in which we describe the ordinary material universe (Jung's pleroma) and the way in which we are forced to describe mind. The contrast lies in this: that, for the material universe, we shall commonly be able to say that the "cause" of an event is some force or impact exerted upon some part of the material system by some one other part. One part acts upon another part. In contrast, in the world of ideas, it takes a *relationship*, either between two parts or between a part at time 1 and the same part at time 2, to activate some third component which we may call the *receiver*.

---

*Philosophie Zoologique (1809), first edition, especially Part III, Chapter 1. Lamarck's title page is here reproduced and a translation follows:
Zoological Philosophy or Exposition of Considerations relative to the natural history of Animals, the diversity of their [internal] organization and of the [mental] faculties which they get from that [organization]; and relative to the physical causes which maintain life in them and give space to the movements which they execute; and finally, relative to those [physical causes] which produce, some of them the perception and others the intelligence of those [animals] which are endowed with those [faculties].

The reader will note that even on his title page Lamarck is careful to insist upon an exact and articulate statement of relations between "physical cause," "organization," "sentiment" and "intelligence." (The translation of the French words, *sentiment* and *intelligence,* is difficult. As I read it, *sentiment* is close to what English speaking psychologists would call "perception," and *intelligence* is close to what we would call "intellect.")

PHILOSOPHIE
ZOOLOGIQUE,
ou
EXPOSITION

Des Considérations relatives à l'histoire naturelle
des Animaux ; à la diversité de leur organisation
et des facultés qu'ils en obtiennent ; aux causes
physiques qui maintiennent en eux la vie et
donnent lieu aux mouvemens qu'ils exécutent ;
enfin , à celles qui produisent , les unes le sentiment , et les autres l'intelligence de ceux qui en
sont doués ;

Par J.-B.-P.-A. LAMARCK,

Professeur de Zoologie au Muséum d'Histoire Naturelle , Membre de
l'Institut de France et de la Légion d'Honneur, de la Société Phi-
lomatique de Paris , de celle des Naturalistes de Moscou , Membre
correspondant de l'Académie Royale des Sciences de Munich, de
la Société des Amis de la Nature de Berlin , de la Société Médicale
d'Emulation de Bordeaux, de celle d'Agriculture, Sciences et Arts
de Strasbourg , de celle d'Agriculture du département de l'Oise ,
de celle d'Agriculture de Lyon , Associé libre de la Société des
Pharmaciens de Paris , etc.

TOME PREMIER.

A PARIS,

Chez { DENTU , Libraire, rue du Pont de Lodi , N°. 3 ;
L'AUTEUR , au Muséum d'Histoire Naturelle (Jardin
des Plantes).

M. DCCC. IX.

What the receiver (e.g., a sensory end organ) responds to is a *difference* or a *change.*

In Jung's pleroma, there are no differences, no distinctions. It is that nonmental realm of description where *difference* between two parts need never be evoked to explain the response of a third.

It is surprising to find how rare are cases in the nonorganic world in which some A responds to a *difference* between some B and some C. The best example I can think of is the case of an automobile traveling over a bump in the road. This instance comes close, at least, to meeting our verbal definition of what happens in processes of perception by mind. External to the automobile there are the two components of a difference: the level of the road and the level of the top of the bump. The car approaches these with its own energy of motion and jumps into the air under impact of the difference, using its own energy for this response. This example contains a number of features closely reminiscent of what happens when a sense organ responds to or collects a piece of information.

The sense of touch is one of the most primitive and simple of the senses, and what sensory information is can easily be illustrated by using touch as an example. In lecturing, I commonly make a heavy dot with chalk on the surface of the blackboard, crushing the chalk a little against the board to achieve some thickness in the patch. I now have on the board something rather like the bump in the road. If I lower my fingertip—a touch-sensitive area—vertically onto the white spot, I shall not feel it. But if I move my finger across the spot, the difference in levels is very conspicuous. I know exactly where the edge of the dot is, how steep it is, and so on. (All this assumes that I have correct opinions about the localization and sensitivity of my fingertip, for many ancillary sorts of information are also needed.)

What happens is that a static, unchanging state of affairs, existing, supposedly, in the outside universe quite regardless of whether we sense it or not, becomes the cause of an event, a step function, a sharp change in the state of the relationship between my fingertip and the surface of the blackboard. My finger goes smoothly over the unchanged surface until I encounter the edge of the white spot. At that moment *in time,* there is a discontinuity, a step; and soon after, there is a reverse step as my finger leaves the spot behind.

This example, which is typical of all sensory experience, shows how our sensory system—and surely the sensory systems of all other creatures (even plants?) and the mental systems behind the senses (i.e., those parts of the mental systems inside the creatures)—can only operate with *events,* which we can call *changes.*

The unchanging is imperceptible unless we are willing to move relative to it.

In the case of vision, it is true that we think we can see the unchanging. We see what looks like the stationary, unmarked blackboard, not just the outlines of the spot. But the truth of the matter is that we continuously do with the eye what I was doing with my fingertip. The eyeball has a continual tremor, called *micronystagmus.* The eyeball vibrates through a few seconds of arc and thereby causes the optical image on the retina to move relative to the rods and cones which are the sensitive end organs. The end organs are thus in continual receipt of events that correspond to *outlines* in the visible world. We *draw* distinctions; that is, we pull them out. Those distinctions that remain undrawn are *not.* They are lost forever with the sound of the falling tree which Bishop Berkeley did not hear.* They are part of William Blake's "corporeal": "Nobody knows of its Dwelling Place: it is in Fallacy, and its Existence an Imposture."†

Notoriously it is very difficult to detect gradual change because along with our high sensitivity to rapid change goes also the phenomenon of accommodation. Organisms become habituated. To distinguish between slow change and the (imperceptible) unchanging, we require information of a different sort; we need a clock.

The matter becomes even worse when we try to judge the *trend-*

---

* The bishop argued that only the perceived is "real" and that the tree which falls *unheard* makes no sound. I would argue that latent differences, i.e., those which for whatever reason do not make a difference, are not *information,* and that "parts," "wholes," "trees," and "sounds" exist as such only in quotation marks. It is *we* who differentiate "tree" from "air," "whole" from "part," and so on. But do not forget that the "tree" is alive and therefore itself capable of receiving certain sorts of information. It too may discriminate "wet" from "dry."

In this book I have many times used quotation marks to remind the reader of these truths. Strictly speaking, every word in the book should be in quotation marks, thus: *"cogito" "ergo" "sum."*
† *Catalogue for the Year 1810.* Blake says elsewhere, "Wise men see outlines and therefore they draw them." He is using the word *draw* in a different sense from that in which we say we "draw" distinctions, but he draws similar conclusions. Attneave has demonstrated that information (i.e., perceptible difference or distinction) is necessarily concentrated at outlines. See Frederick Attneave, *Applications of Information Theory to Psychology* (New York: Holt, Rinehart and Winston, 1959).

ing of phenomena that are characteristically changeable. The weather, for example, is continually changing—from hour to hour, from day to day, from week to week. But is it changing from year to year? Some years are wetter and some hotter, but is there a trend in this continual zigzag? Only statistical study, over periods longer than human memory, can tell us. In such cases we need information about *classes* of years.

Similarly, it is very difficult for us to perceive changes in our own social affairs, in the ecology around us, and so on. How many people are conscious of the astonishing decrease in the number of butterflies in our gardens? Or of birds? These things undergo drastic change, but we become accustomed to the new state of affairs before our senses can tell us that it is new.

The feinting of a boxer, who makes moves as if to hit with his left hand without hitting, deceives us into believing that that left hand is not going to hit—until it does hit, and we are unpleasantly surprised.

It is a nontrivial matter that we are almost always unaware of trends in our changes of state. There is a quasi-scientific fable that if you can get a frog to sit quietly in a saucepan of cold water, and if you then raise the temperature of the water very slowly and smoothly so that there is no moment *marked* to be the moment at which the frog should jump, he will never jump. He will get boiled. Is the human species changing its own environment with slowly increasing pollution and rotting its mind with slowly deteriorating religion and education in such a saucepan?

But I am concerned at this moment only with understanding how mind and mental process must *necessarily* work. What are their limitations? And, precisely because the mind can receive news only of difference, there is a difficulty in discriminating between a *slow change* and a *state*. There is necessarily a threshold of gradient below which gradient cannot be perceived.

Difference, being of the nature of relationship, is not located in time or in space. We say that the white spot is "there," "in the middle of the blackboard," but the difference between the spot and the blackboard is not "there." It is not in the spot; it is not in the blackboard; it is not in the space between the board and the chalk. I could perhaps lift the chalk off the board and send it to Australia, but the difference would not be destroyed or even shifted because difference does not have location.

When I wipe the blackboard, where does the difference go? In one sense, the difference is randomized and irreversibly gone, as "I" shall be gone when I die. In another sense, the difference will endure as an idea—as part of my *karma*—as long as this book is read, perhaps as long as the ideas in this book go on to form other ideas, reincorporated into other minds. But this enduring karmic information will be information about an imaginary spot on an imaginary blackboard.

Kant argued long ago that this piece of chalk contains a million potential facts (*Tatsachen*) but that only a very few of these become truly facts by affecting the behavior of entities capable of responding to facts. For Kant's *Tatsachen,* I would substitute *differences* and point out that the number of *potential* differences in this chalk is infinite but that very few of them become *effective* differences (i.e., items of information) in the mental process of any larger entity. *Information* consists of differences that make a difference.

If I call attention to the difference between the chalk and a piece of cheese, you will be affected by that difference, perhaps avoiding the eating of the chalk, perhaps tasting it to verify my claim. Its noncheese nature has become an effective difference. But a million other differences—positive and negative, internal and external to the chalk—remain latent and ineffective.

Bishop Berkeley was right, at least in asserting that what happens in the forest is *meaningless* if he is not there to be affected by it.

We are discussing a world of *meaning,* a world some of whose details and differences, big and small, in some parts of that world, get *represented* in relations between other parts of that total world. A change in my neurons or in yours must represent that change in the forest, that falling of that tree. But not the physical event, only the *idea* of the physical event. And the idea has no location in space or time—only perhaps in an *idea* of space or time.

Then there is the concept "energy," whose precise referent is fashionably concealed by contemporary forms of obscurantism. I am not a physicist, not up to date in modern physics, but I note that there are two conventional definitions or *aspects* (is that the word?) of "energy." I have a difficulty in understanding these two definitions simultaneously—they seem to conflict. But it is clear to me that neither definition is relevant to what I am talking about.

One definition asserts that "energy" is of the same order of abstraction as "matter"; that both are somehow *substances* and are mutually convertible one into the other. But difference is precisely *not* substance.

The other definition is more old-fashioned and describes energy as having the dimensions MV^2. Of course, difference, which is usually a *ratio* between similars, has no dimensions. It is *qualitative,* not *quantitative.* (See Chapter 2, in which the relation between quantity and quality or pattern was examined.)

For me, the word *stimulus* denotes a member of a class of information coming in through a sense organ. For many speakers, it seems to mean a push or shot of "energy."

If there are readers who still want to equate information and difference with energy, I would remind them that *zero* differs from *one* and can therefore trigger response. The starving amoeba will become more active, hunting for food; the growing plant will bend away from the dark, and the income tax people will become alerted by the declarations which you did not send. Events *which are not* are different from those which might have been, and events which are not surely contribute no energy.

CRITERION 3. MENTAL PROCESS REQUIRES COLLATERAL ENERGY

Although it is clear that mental processes are triggered by difference (at the simplest level) and that difference is *not* energy and usually contains no energy, it remains necessary to discuss the energetics of mental process because processes, of whatever kind, require energy.

Living things are subject to the great conservative regularities of physics. The laws of conservation of mass and of energy apply completely to living creatures. There is no creation or destruction of energy (MV^2) in the business of living. On the other hand, the *syntax* for the describing of the energetics of life is a different syntax from that which was used 100 years ago to describe the energetics of force and impact. This difference of syntax is my third criterion of mental process.

There is a tendency today among subatomic physicists to use

metaphors taken from life to describe the events inside the accelerator. No doubt this trick of speech, technically called the *pathetic fallacy,* is as wrong as that of which I complain, although less dangerous. To liken the mountain to a man and talk of its "humor" or "rage" does little harm. But to liken the man to the mountain proposes that all human relationships are what Martin Buber might call *I-it* or perhaps *it-it* relations. The mountain, personified in our speech, will not become a person, will not *learn* a more personal way of being. But the human being, depersonified in his own talk and thought, may indeed learn more thingish habits of action.

In the opening paragraph of this section, the word *triggered* was used with intent. The metaphor is not perfect,* but it is at least more appropriate than all the metaphoric forms which ascribe relevance to the energy contained in the stimulus event. Billiard-ball physics proposes that when ball A hits ball B, A *gives* energy to B, which responds *using* this energy which A gave it. That is the old syntax and is profoundly, deeply nonsense. Between billiard balls, there is, of course, no "hitting" or "giving" or "responding" or "using." Those words come out of the habit of personifying things and, I suppose, make it easier to go from that nonsense to thingifying people—so that when we speak of the "response" of a living thing to an "external stimulus," we seem to be talking about something like what happens to a billiard ball when it is hit by another.

When I kick a stone, I give energy to the stone, and it moves with that energy; and when I kick a dog, it is true that my kick has a partly Newtonian effect. If it is hard enough, my kick might put the dog into Newtonian orbit, but that is not the essence of the matter. When I kick a dog, it responds with energy got from metabolism. In the "control" of action by information, the energy is already available in the respondent, in advance of the impact of events.

The trick, which life plays continually but which undomes-

---

* Firearms are a somewhat inappropriate metaphor because in most simple firearms, there is only a lineal sequence of energetic dependencies. The trigger releases a pin or hammer whose movement, when released, is energized by a spring. The hammer fires a percussion cap which is energized by chemical energy to provide an intense exothermic reaction, which sets alight the main supply of explosive in the cartridge. In *nonrepeating* firearms, the marksman must now replace the energetic chain, inserting a new cartridge with new percussion cap. In biological systems, the end of the lineal sequence sets up conditions for a future repetition.

ticated matter plays only rarely, is familiar. It is the trick of the faucet, the switch, the relay, the chain reaction, and so on—to name a few instances in which the nonliving world does indeed simulate true living in a gross way.

In all these cases, the energy for the response or effect was available in the respondent before the event occurred which triggered it. The kids who say they are "turned on" by certain experiences of sight or sound are using a metaphor which almost makes sense. They would do better still if they said that the music or the pretty face "released" them.

In life and its affairs, there are typically two energetic systems in interdependence: One is the system that uses its energy to open or close the faucet or gate or relay; the other is the system whose energy "flows through" the faucet or gate when it is open.

The ON position of the switch is a pathway for the passage of energy which originates elsewhere. When I turn the faucet, my work in turning the faucet does not push or pull the flow of the water. That work is done by pumps or gravity whose force is set free by my opening the faucet. I, in "control" of the faucet, am "permissive" or "constraining"; the flow of the water is energized from other sources. I partly determine what pathways the water will take if it flows at all. *Whether* it flows is not my immediate business.

The combining of the two systems (the machinery of decision and the source of energy) makes the total relationship into one of partial mobility on each side. You can take a horse to the water, but you cannot make him drink. The drinking is his business. But even if your horse is thirsty, he cannot drink unless you take him. The taking is your business.

But I oversimplify the matter by focusing only on the energetics. There is also the generalization (criterion 2) that only difference can trigger response. We have to combine that generalization with what has just been said about the typical relation of energy sources and with the remaining criteria of mental process, namely, the organization of triggered events into circuits, coding, and the genesis of hierarchies of meaning.

CRITERION 4. MENTAL PROCESS REQUIRES CIRCULAR (OR MORE COMPLEX) CHAINS OF DETERMINATION

If mere survival, mere continuance, is of interest, then the harder sorts of rocks, such as granite, have to be put near the top of the list as most successful among macroscopic entities. They have retained their characteristics unchanged since quite early in the formation of the earth's crust and have achieved this in many varied environments from poles to tropics. If the simple tautology of the theory of natural selection be stated as "those descriptive propositions which remain true for longest time remain true longer than those that become untrue sooner," then granite is a more successful entity than any species of organism.

But the rock's way of staying in the game is different from the way of living things. The rock, we may say, *resists* change; it stays put, unchanging. The living thing escapes change either by correcting change or changing itself to meet the change or by incorporating continual change into its own being. "Stability" may be achieved either by rigidity or by continual repetition of some cycle of smaller changes, which cycle will return to a *status quo ante* after every disturbance. Nature avoids (temporarily) what looks like irreversible change by accepting ephemeral change. "The bamboo bends before the wind," in Japanese metaphor; and death itself is avoided by a quick change from individual subject to class. Nature, to personify the system, allows old man Death (also personified) to have his individual victims while she substitutes that more abstract entity, the class or taxon, to kill which Death must work faster than the reproductive systems of the creatures. Finally, if Death should have his victory over the species, Nature will say, "Just what I needed for my ecosystem."

All this becomes possible by combination of those criteria of mental process that have already been mentioned with this fourth criterion, that the organization of living things depends upon circular and more complex chains of determination. All the fundamental criteria are combined to achieve success in that mode of survival which characterizes life.

The idea that circular causation is of very great importance was first generalized at the end of World War II by Norbert Wiener and

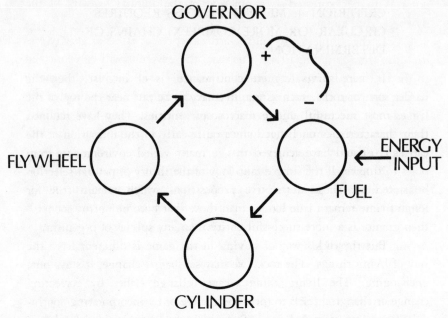

Figure 8

perhaps other engineers who were working with the mathematics of nonliving systems (i.e., machines). This matter is best understood by means of a highly simplified mechanical diagram (Figure 8).

Imagine a machine in which we distinguish, say, four parts, which I have loosely called "flywheel," "governor," "fuel," and "cylinder." In addition, the machine is connected to the outside world in two ways, "energy input" and "load," which is to be imagined as variable and perhaps weighing upon the flywheel. The machine is circular in the sense that flywheel drives governor which alters fuel supply which feeds cylinder which, in turn, drives flywheel.

Because the system is circular, effects of events at any point in the circuit can be carried all around to produce changes at that point of origin.

In such a diagram, arrows are used to indicate direction from cause to effect, and it is possible to imagine any combination of types of causation from step to step. The arrows may be supposed to represent mathematical functions or equations showing the *types of effect* that successive parts have on each other. Thus, the angle of the arms of the governor is to be expressed as a function of the angular velocity of the flywheel. And so on.

In the simplest case, all the arrows represent either *no gain* or *positive gain* from part to part. In this case, the governor will be connected to the fuel supply in a way which no engineer would approve, namely, so that the more the arms of the governor diverge, the more the fuel. So rigged, the machine will go into a *runaway,* operating exponentially faster and faster, until either some part breaks or perhaps the fuel duct can deliver fuel at no greater rate.

But the system might equally be set up with one or more inverse relations at arrow junctures. This is the usual way of setting up governors, and the name *governor* is applied to that part which provides the first half of such a relation. In this case, the *more* the arms diverge, the *less* the fuel supply.

As a matter of history, systems with positive gain, variously called *escalating* or *vicious* circles, were anciently familiar. In my own work with the Iatmul tribe on the Sepik River in New Guinea, I had found that various relations among groups and among various types of kin were characterized by interchanges of behavior such that the more A exhibited a given behavior, the more B was likely to exhibit the same behavior. These I called *symmetrical* interchanges. Conversely, there were also stylized interchanges in which B's behavior was different from, but *complementary* to, that of A. In either case, the relations were potentially subject to progressive escalation, which I called *schismogenesis.*

I noted at that time that either symmetrical or complementary schismogenesis could conceivably lead to runaway and the breakdown of the system. There was positive gain at each interchange and a sufficient supply of energy from the metabolism of the persons concerned to destroy the system in rage or greed or shame. It takes rather little energy (MV^2) to enable a human being to destroy others or the integration of a society.

In other words, in the 1930s I was already familiar with the idea of "runaway" and was already engaged in classifying such phenomena and even speculating about possible combinations of different sorts of runaway. But at that time, I had no idea that there might be circuits of causation which would contain one or more negative links and might therefore be self-corrective. Nor, of course, did I see that runaway systems, such as population growth, might contain the seeds of their own self-correction in the form of epidemics, wars, and government programs.

Many self-corrective systems were also already known. That is, individual cases were known, but the *principle* remained unknown. Indeed, occidental man's repeated discovery of instances and inability to perceive the underlying principle demonstrate the rigidity of his epistemology. Discoveries and rediscoveries of the principle include Lamarck's transformism (1809), James Watt's invention of the governor for the steam engine (late eighteenth century), Alfred Russel Wallace's perception of natural selection (1856), Clark Maxwell's mathematical analysis of the steam engine with a governor (1868), Claude Bernard's *milieu interne,* Hegelian and Marxian analyses of social process, Walter Cannon's *Wisdom of the Body* (1932), and the various mutually independent steps in the development of cybernetics and systems theory during and immediately after World War II.

Finally, the famous paper in *Philosophy of Science* by Rosenblueth, Wiener, and Bigelow* proposed that the self-corrective circuit and its many variants provided possibilities for modeling the adaptive actions of organisms. The central problem of Greek philosophy—the problem of purpose, unsolved for 2,500 years—came within range of rigorous analysis. It was possible to model even such marvelous sequences as the cat's jump, timed and directed to land where the mouse will be when the cat lands.

In passing, however, it is worth asking whether the difficulty in recognizing this basic cybernetic principle was due only to humankind's laziness when asked to make a basic change in the paradigms of its thought or whether there were other processes preventing acceptance of what seems to have been, as we look back, a very simple idea. Was the older epistemology itself reinforced by self-corrective or runaway circuits?

A rather detailed account of the nineteenth-century history of the steam engine with governor may help the reader to understand both the circuits and the blindness of the inventors. Some sort of governor was added to the early steam engine, but the engineers ran into difficulties.

* Rosenblueth, A., N. Wiener, and J. Bigelow, "Behavior, Purpose and Teleology," *Philosophy of Science* 10 (1943): 18–24.

They came to Clark Maxwell with the complaint that they could not draw a blueprint for an engine with a governor. They had no theoretical base from which to predict how the machine that they had drawn would behave when built and running.

There were several possible sorts of behavior: Some machines went into runaway, exponentially maximizing their speed until they broke or slowing down until they stopped. Others oscillated and seemed unable to settle to any mean. Others—still worse—embarked on sequences of behavior in which the amplitude of their oscillation would itself oscillate or would become greater and greater.

Maxwell examined the problem. He wrote out formal equations for relations between the variables at each successive step around the circuit. He found, as the engineers had found, that combining this set of equations would not solve the problem. Finally, he found that the engineers were at fault in not considering *time*. Every given system embodied relations to time, that is, was characterized by time constants determined by the given *whole*. These constants were not determined by the equations of relationship between successive parts but were *emergent* properties of the system.

Imagine for a moment that the engine is running smoothly and encounters a load. It must go uphill or drive some appliance. Immediately, the angular velocity of the flywheel will fall off. This will cause the governor to spin less fast. The weighted arms of the governor will fall, reducing the angle between arms and shaft. As this angle decreases, more fuel will be injected into the cylinder, and the machine will speed up, changing the angular velocity of the flywheel in a sense contrary to that change which the load had induced.

But whether the corrective change will precisely correct the changes that the load induced is a question of some difficulty. After all, the whole process occurs in time. At some time 1, the load was encountered. The change in the speed of the flywheel *followed* time 1. The changes in the governor followed still later. Finally the corrective message reached the flywheel at some time 2, later than time 1. But the amount of the correction was determined by the amount of deviation at time 1. By time 2, the deviation will have changed.

At this point, note that a very interesting phenomenon has occurred within our description of the events. When we were talking as if we were inside the circuit, we noted changes in the behavior of the parts whose magnitude and timing were determined by forces and impacts between the separate components of the circuit. Step by step around the circuit, my language had the general form: A change in A determines a change in B. And so on. But when the description reaches the place from which it (arbitrarily) started, there is a sudden change in this syntax. The description now must compare change with change and use the result of *that* comparison to account for the next step.

In other words, a subtle change has occurred in the subject of discourse, which, in the jargon of the last section (criterion 6) of this chapter, we shall call a change in *logical typing*. It is the difference between talking in a language which a physicist might use to describe how one variable acts upon another and talking in another language about the circuit as a whole which reduces or increases difference. When we say that the system exhibits "steady state" (i.e., that in spite of variation, it retains a median value), we are talking about the circuit as a whole, not about the variations within it. Similarly the question which the engineers brought to Clark Maxwell was about the circuit as a whole: How can we plan it to achieve a steady state? They expected the answer to be in terms of relations between the individual variables. What was needed and supplied by Maxwell was an answer in terms of the time constants of the total circuit. This was the bridge between the two levels of discourse.

The entities and variables that fill the stage at one level of discourse vanish into the background at the next-higher or -lower level. This may be conveniently illustrated by considering the referent of the word *switch,* which engineers at times call a *gate* or *relay*. What goes through is energized from a source that is different from the energy source which opens the gate.

At first thought a "switch" is a small contraption on the wall which turns the light on or off. Or, with more pedantry, we note that the light is turned on or off by human hands "using" the switch. And so on.

We do not notice that the concept "switch" is of quite a different order from the concepts "stone," "table," and the like. Closer examination shows that the switch, considered as a part of an electric circuit, *does not exist* when it is in the on position. From the point of view of the circuit, it is not different from the conducting wire which leads to it and the wire which leads away from it. It is merely "more conductor." Conversely, but similarly, when the switch is off, it does not exist from the point of view of the circuit. It is nothing, a gap between two conductors which, themselves exist only as conductors when the switch is on.

In other words, the switch is *not* except at the moments of its change of setting, and the concept "switch" has thus a special relation to *time*. It is related to the notion "change" rather than to the notion "object."

Sense organs, as we have already noted, admit only news of difference and are indeed normally triggered only by change, i.e., by events or by those differences in the perceived world which can be made into events by moving the sense organ. In other words, the end organs of sense are analogous to switches. They must be turned "on" for a single moment by external impact. That single moment is the generating of a single impulse in the afferent nerve. The threshold (i.e., the amount of event required to throw the switch) is, of course, another matter and may be changed by many physiological circumstances, including the state of the neighboring end organs.

The truth of the matter is that every circuit of causation in the whole of biology, in our physiology, in our thinking, our neural processes, in our homeostasis, and in the ecological and cultural systems of which we are parts—every such circuit conceals or proposes those paradoxes and confusions that accompany errors and distortions in logical typing. This matter, closely tied both to the matter of circuitry and to the matter of coding (criterion 5), will be considered more fully in the discussion of criterion 6.

CRITERION 5. IN MENTAL PROCESS, THE EFFECTS OF DIFFERENCE ARE TO BE REGARDED AS TRANSFORMS (i.e., CODED VERSIONS) OF THE DIFFERENCE WHICH PRECEDED THEM

At this point, we must consider how the differences examined in the discussion of criterion 2 and their trains of effect in promoting other differences become material of information, redundancy, pattern, and so on. First, we have to note that any object, event, or difference in the so-called "outside world" can become a source of information provided that it is incorporated into a circuit with an appropriate network of flexible material in which it can produce changes. In this sense, the solar eclipse, the print of the horse's hoof, the shape of the leaf, the eyespot on a peacock's feather—whatever it may be—can be incorporated into mind if it touches off such trains of consequence.

We proceed, then, to the broadest-possible statement of Korzybski's famous generalization. He asserted that *the map is not the territory*. Looking at the matter in the very wide perspective that we are now using, we see the map as some sort of effect summating differences, organizing news of differences in the "territory." Korzybski's map is a convenient metaphor and has helped a great many people, but boiled down to its ultimate simplicity, his generalization asserts that the effect is not the cause.

This—the fact of difference between effect and cause when both are incorporated into an appropriately flexible system—is the primary premise of what we may call *transformation* or *coding*.

Some regularity in the relation between effect and cause is, of course, assumed. Without that, no mind could possibly guess at cause from effect. But granted such a regularity, we can go on to classify the various sorts of relationship that can obtain between effect and cause. This classification will later embrace very complex cases when we encounter complex aggregates of information that may be called *patterns, action sequences,* and the like.

Even greater variety of transformation or coding arises from the fact that the respondent to difference is almost universally energized by collateral energy. (Criterion 3, above.) There then need be no simple relation between the magnitude of the event or difference which triggers the response and the resulting response.

However, the first dichotomy that I wish to impose on the multitudinous varieties of transformation is that which would divide the cases in which response is *graded* according to some variable in the trigger event, as opposed to those in which the response is a matter of on-off thresholds. The steam engine with a governor provides a typical instance of one type, in which the angle of the arms of the governor is continuously variable and has a continuously variable effect on the fuel supply.

In contrast, the house thermostat is an on-off mechanism in which temperature causes a thermometer to throw a switch at a certain level. This is the dichotomy between *analogic* systems (those that vary continuously and in step with magnitudes in the trigger event) and *digital* systems (those that have the on-off characteristic).

Notice that the digital systems more closely resemble systems containing number; whereas analogic systems seem to be dependent more on quantity. The difference between these two genera of coding is an example of the generalization (discussed in Chapter 2) that number is different from quantity. There is a discontinuity between each number and the next, as in digital systems there is discontinuity between "response" and "no response." This is the discontinuity between "yes" and "no."

In the early days of cybernetics, we used to argue about whether the brain is, on the whole, an analogic or a digital mechanism. That argument has since disappeared with the realization that description of the brain has to start from the all-or-nothing characteristic of the neuron. At least in a vast majority of instances, the neuron either fires or does not fire; and if this were the end of the story, the system would be purely digital and binary. But it is possible to make systems out of digital neurons that will have the *appearance* of being analogic systems. This is done by the simple device of multiplying the pathways so that a given cluster of pathways might consist of hundreds of neurons, of which a certain percentage would be firing and a certain other percentage would be quiet, thus giving an apparently graded response. In addition, the individual neuron is modified by hormonal and other environmental conditions around it that may alter its threshold in a truly quantitative manner.

I recall, however, that in those days, before we had fully realized the degree to which analogical and digital characteristics might be combined in one system, the discussants who argued to and fro on the question of whether the brain is analogic or digital showed very marked individual and irrational preferences for one or the other view. I tended to prefer hypotheses stressing the digital; whereas those more influenced by physiology and perhaps less by the phenomena of language and overt behavior tended to favor the analogic explanations.

Other classifications of types of coding are important in the problem of recognizing mental characteristics in very primitive entities. In some highly diffuse systems, it is not easy, perhaps not possible, to recognize either sense organs or pathways along which information travels. Ecosystems such as a seashore or a redwood forest are undoubtedly self-corrective. If in a given year the population of some species is unusually increased or reduced, within a very few years that population will return to its usual level. But it is not easy to point to any part of the system which is the sense organ gathering information and influencing corrective action. I think that such systems are quantitative and gradual and that the quantities whose *differences* are the informational indicators are at the same time quantities of needed supplies (food, energy, water, sunlight, and so on). A great deal of research has been done on the energy pathways (e.g., food chains and water supplies) in such systems. But I do not know of any specific study that looks at these supplies as carrying immanent information. It would be nice to know whether these are analogic systems in which *difference* between events in one round of the circuit and events in the next round (as in the steam engine with governor) becomes the crucial factor in the self-corrective process.

When the growing seedling bends toward light, it is influenced by difference in illumination and grows more rapidly on the darker side, thus bending and catching more light—a substitute for locomotion depending upon difference.

Two other forms of transform or coding are worth mentioning because they are very simple and very easily overlooked. One is *template* coding, in which, for example, in the growth of any organism, the shape and morphogenesis that occur at the growing point are commonly defined by the state of the growing surface at the time of growth. To cite a very trivial example, the trunk of a palm tree continues more or less parallel-sided from the bole up to the top, where the growing point is. At any point, the growing tissue, or cambium, is depositing wood downward behind it on the face of the already grown trunk. That is, the shape of what it deposits is determined by the shape of the previous growth. Similarly, in regeneration of wounds and such things, it would seem that rather often the shape of the regenerative tissue and its differentiation are determined by the shape and differentiation of the cut face. This is perhaps as near to a case of "direct" communication as can be imagined. But it should be noted that in many cases, the growth of, for example, the regenerating organ has to be the mirror image of the state of affairs at the interface with the old body. If the face is indeed two-dimensional and has no depth, then the growing component presumably takes its depth direction from some other source.

The other type of communication that is often forgotten is called *ostensive*. If I say to you, "That's what a cat looks like," pointing to the cat, I am using the cat as an ostensive component in my communication. If I walk down the street and see you coming and say, "Oh, there's Bill," I have received information ostensively from you, (your appearance, your walk, and so on) whether or not you intended to transmit that information.

Ostensive communication is peculiarly important in language learning. Imagine a situation in which a speaker of a given language must teach that language to some other individual under circumstances in which ostensive communication is strictly limited. Suppose A must teach B a language totally unknown to B over the telephone and that they have no other language in common. A will be able, perhaps, to communicate to B some characteristics of voice, of cadence, even of grammar; but it is quite impossible for A to tell B what any word "means" in the ordinary sense. So far as B is concerned, substantives and verbs will be only grammatical entities, not names of identifiable objects. Cadence, sequential structure, and the like are present in the sequence of sounds sent over the telephone and can conceivably be "pointed to" and therefore taught to B.

Ostensive communication is perhaps similarly necessary in the learning of any transformation or code. For example, in all learning experiments, the giving or withholding of the reinforcement is an approximate method of pointing to the right response. In the training of performing animals, various devices are used to make this pointing more accurate. The trainer may have a whistle that is very briefly tooted at the precise moment when the animal does the right thing, thereby using the responses of the learner as ostensive examples in the teaching.

Another form of very primitive coding which is ostensive is *part-for-whole* coding. For example, I see a redwood tree standing up out of the ground, and I know from this perception that underneath the ground at that point I shall find roots, or I hear the beginnings of a sentence and know at once from that beginning the grammatical structure of the rest of the sentence and may very well know many of the words and ideas contained in it. We live in a life in which our percepts are perhaps always the perception of *parts,* and our guesses about wholes are continually being verified or contradicted by the later presentation of other parts. It is perhaps so, that *wholes* can never be presented; for *that* would involve direct communication.

CRITERION 6. THE DESCRIPTION AND CLASSIFICATION OF THESE PROCESSES OF TRANSFORMATION DISCLOSES A HIERARCHY OF LOGICAL TYPES IMMANENT IN THE PHENOMENA

This section must undertake two tasks: first, to make the reader understand what is meant by logical types and related ideas, which, in various forms, have fascinated man for at least 3,000 years. Second, to persuade the reader that what I am talking about is characteristic of mental process and is even a necessary characteristic. Neither of these two tasks is entirely simple, but William Blake commented, "Truth can never be told so as to be understood and not be believ'd." So, the two tasks become one task, that of exhibiting the truth so that it can be understood; though I well know that to tell the truth in any important area of life so as to be understood is an excessively difficult feat, in which Blake himself rarely succeeded.

I shall begin with an abstract presentation of what I mean, and I shall follow that with rather simple cases to illustrate the ideas. Finally, I shall try to drive home the importance of this criterion by exhibiting cases in which the discrimination of levels of communication has been so confused or distorted that various sorts of frustration and pathology have been the result.

For the abstract presentation, consider the case of a very simple relationship between two organisms in which organism A has emitted some sort of sound or posture from which B could learn something about the state of A relevant to B's own existence. It might be a threat, a sexual advance, a move towards nurturing, or an indication of membership in the same species. I already noted in the discussion of coding (criterion 5) that no message, under any circumstances, is that which precipitated it. There is always a partly predictable and therefore rather regular relation between message and referent, that relation indeed never being direct or simple. Therefore, if B is going to deal with A's indication, it is absolutely necessary that B know what those indications mean. Thus, there comes into existence another *class* of information, which B must assimilate, to tell B about the coding of messages or indications coming from A. Messages of this class will be, not about A or B, but about the coding of messages. They will be of a different logical type. I will call them *metamessages*.

Again, beyond messages about simple coding, there are much more subtle messages that become necessary because codes are conditional; that is, the meaning of a given type of action or sound changes relative to *context,* and especially relative to the changing state of the relationship between A and B. If at a given moment the relation becomes playful, this will change the meaning of many signals. It was the observation that this was true for the animal as well as the human world which led me to the research that generated the so-called *double bind* theory of schizophrenia and to the whole epistemology offered in this book. The zebra may identify (for the lion) the nature of the context in which they meet by bolting, and even the well-fed lion may give chase. But the hungry lion needs no such labeling of that particular context. He learned long ago that zebras can be eaten. Or was this lesson so early as to require no teaching? Were parts of the necessary knowledge innate?

The whole matter of messages which make some other message intelligible by putting it in context must be considered, but in the *absence* of such metacommunicative messages, there is still the possibility that B will ascribe context to A's signal, being guided in this by genetic mechanisms.

It is perhaps at this abstract level that learning and genetics meet. Genes may perhaps influence an animal by determining how it will perceive and classify the contexts of its learning. But mammals, at least, are capable also of *learning about context.*

What used to be called *character*—i.e., the system of interpretations which we place on the contexts we encounter—can be shaped both by genetics and by learning.

All this is premised on the existence of *levels* whose nature I am here trying to make clear. We start, then, with a potential differentiation between action in context and action or behavior which defines context or makes context intelligible. For a long time, I referred to the latter type of communication as *metacommunication,* borrowing this term from Whorf.*

A function, an effect, of the metamessage is in fact to *classify* the messages that occur within its context. It is at this point that the theory offered here connects with the work of Russell and Whitehead in the first ten years of this century, finally published in 1910 as *Principia Mathematica.*† What Russell and Whitehead were tackling was a very abstract problem. Logic, in which they believed, was to be salvaged from the tangles created when the *logical types,* as Russell called them, are maltreated in mathematical presentation. Whether Russell and Whitehead had any idea when they were working on *Principia* that the matter of their interest was vital to the life of human beings and other organisms, I do not know. Whitehead certainly knew that human beings could be amused and humor generated by kidding around with the types. But I doubt whether he ever made the step from enjoying this game to seeing that the game was nontrivial and would cast light on the whole of biology. The more general insight was—perhaps unconsciously—avoided rather than contemplate the nature of the human dilemmas that the insight would propose.

The mere fact of humor in human relations indicates that at least at this biological level, multiple typing is essential to human communication. In the absence of the distortions of logical typing, humor would be unnecessary and perhaps could not exist.

Even at a very abstract level, phenomena provoked by logical typing have fascinated thinkers and fools for many thousands of years. But logic had to be saved from the paradoxes which clowns might enjoy. One of the first things that Russell and Whitehead observed in attempting this was that the ancient paradox of Epimenides—"Epimenides was a Cretan who said, 'Cretans always lie' "—was built upon classification and metaclassification. I have presented the paradox here in the form of a quotation within a quotation, and this is precisely how the paradox is

* B. L. Whorf, *Language, Thought, and Reality* (Cambridge, Mass.: Technical Press of Massachusetts Institute of Technology, 1956).
† A. N. Whitehead and B. Russell, *Principia Mathematica,* 2d ed. (Cambridge: Cambridge University Press, 1910–1913).

generated. The larger quotation becomes a classifier for the smaller, until the smaller quotation takes over and reclassifies the larger, to create contradiction. When we ask, "Could Epimenides be telling the truth?" the answer is: "If yes, then no," *and* "If no, then yes."

Norbert Wiener used to point out that if you present the Epimenides paradox to a computer, the answer will come out YES . . . NO . . . YES . . . NO . . . until the computer runs out of ink or energy or encounters some other ceiling. As I noted in Chapter 2, section 16, logic cannot model causal systems, and paradox is generated when time is ignored.

If we look at any living organism and start to ask about its actions and postures, we meet with such a tangle or network of messages that the theoretical problems outlined in the previous paragraph become confused. In the enormous mass of interlocking observation, it becomes exceedingly difficult to say that this message or position of the ears is, in fact, *meta-* to that other observation of the folding of the front legs or the position of the tail.

In front of me on the table is a sleeping cat. While I was dictating the last hundred words, the cat changed her position. She was sleeping on her right side, her head pointing more or less away from me, her ears in a position that did not suggest to me alertness, eyes closed, front feet curled up—a familiar arrangement of the body of a cat. While I spoke and, indeed, was watching the cat for behavior, the head turned toward me, the eyes remained closed, respiration changed a little, the ears moved into a half alert position; and it appeared, rightly or wrongly, that the cat was now still asleep but aware of my existence and aware, perhaps, that she was a part of the dictated material. This increase of attention happened *before* the cat was mentioned, that is, before I began to dictate the present paragraph. *Now,* with the cat fully mentioned, the head has gone down, the nose is between the front legs, the ears have stopped being alert. She has decided that her involvement in the conversation does not matter.

Watching this sequence of cat behavior and the sequence of my reading of it (because the system we are talking about is, in the end, not just cat but man-cat and perhaps should be considered more complexly than that, as man-watching-man's-watching-cat-watching-man), there is a hierarchy of contextual components as well as a hierarchy concealed within the enormous number of signals given by the cat about herself.

What seems to be the case is that the messages emanating from the cat are interrelated in a complex net, and the cat herself might be surprised if she could discover how difficult it is to unscramble that mass. No doubt another cat would do the unscrambling better than a human being. But to the human being—and even the trained ethologist is often surprised—the relations between component signals are confused. However, the human "understands" the cat by putting the pieces together *as if* he really knew what is happening. He forms *hypotheses,* and these are continually checked or corrected by less ambiguous actions of the animal.

Cross-species communication is *always* a sequence of contexts of learning in which each species is continually being corrected as to the nature of each previous context.

In other words, the metarelations between particular signals may be confused but understanding may emerge again as true at the next more abstract level.*

In some contexts of animal behavior or relations between human and animal, the levels are in some degree separated not only by the human but also by the animal. I shall exemplify this in two narratives, the first a discussion of the classical Pavlovian experiments on experimental neurosis and the second an account of research into human-dolphin relations with which I was connected at the Oceanic Institute in Hawaii. This will constitute a pair of contrasting cases, in one of which the tangle leads to pathology, while in the other the logical types are finally transcended by the animal.

The Pavlovian case is very famous, but my interpretation of it is different from the standard interpretation, and this difference consists precisely in my insistence on the relevance of context to meaning, which relevance is an example of one set of messages metacommunicative to another.

The paradigm for experimental neurosis is as follows: A dog

---

* The reader is reminded here of what was said about the fallacy of Lamarckism (Chapter 2, section 7). Lamarck proposed that environmental impact could directly affect the genes of the single individual. That is untrue. What is true is a proposition of next-higher logical type: that the environment does have direct impact on the gene pool of the *population*.

(commonly a male) is trained to respond differentially to two alternative "conditioned stimuli," for instance, a circle or an ellipse. In response to X, he is to do A; in response to Y, he is to do B. If in his responses, the dog exhibits this differentiation, he is said to discriminate between the two stimuli and he is positively reinforced or, in Pavlovian language, given an "unconditioned stimulus" of food. When the dog is able to discriminate, the task is made somewhat more difficult by the experimenter, who will either make the ellipse somewhat fatter or make the circle somewhat flatter so that the contrast between the two stimulus objects becomes less. At this point, the dog will have to put out extra effort to discriminate between them. But when the dog succeeds in doing this, the experimenter will again make things more difficult by a similar change. By such a series of steps, the dog is led to a situation in which finally he cannot discriminate between the objects. At this point, if the experiment has been performed with sufficient rigor, the dog will exhibit various symptoms. He may bite his keeper, he may refuse food, he may become disobedient, he may become comatose, and so on. Which set of symptoms the dog exhibits depends, it is claimed, upon the "temperament" of the dog, excitable dogs choosing one set of symptoms and lethargic dogs choosing another.

Now, from the point of view of the present chapter, we have to examine the difference between two verbal forms contained in the orthodox explanation of this sequence. One verbal form is "the dog *discriminates* between the two stimuli"; the other is "the dog's *discrimination* breaks down." In this jump, the scientist has moved from a statement about a particular incident or incidents which can be *seen* to a generalization that is hooked up to an abstraction—"discrimination"—located *beyond vision* perhaps inside the dog. It is this jump in logical type that is the theorist's error. I can, in a sense, see the dog *discriminate,* but I cannot possibly see his "discrimination." There is a jump here from particular to general, from member to class. It seems to me that a better way of saying it would depend upon asking: "What has the dog learned in his training that makes him unable to *accept* failure at the end?" And the answer to this question would seem to be: The dog has learned that *this is a context for discrimination.* That is, that he "should" look for two stimuli and "should" look for the possibility of acting on a difference between them.

For the dog, this is the "task" which has been set—the context in which success will be rewarded.*

Obviously, a context in which there is no perceptible difference between the two stimuli is not one for discrimination. I am sure the experimenter could induce neurosis by using a single object repeatedly and tossing a coin each time to decide whether this single object should be interpreted as an X or as a Y. In other words, an appropriate response for the dog would be to take out a coin, toss it, and use the fall of the coin to decide his action. Unfortunately, the dog has no pocket in which to carry coins and has been very carefully trained in what has now become a lie; that is, the dog has been trained to expect a context for discrimination. He now imposes this interpretation on a context that is not a context for discrimination. He has been taught *not* to discriminate between two classes of contexts. He is in that state *from which the experimenter* started: unable to distinguish contexts.

From the dog's point of view (consciously or unconsciously), to learn context is different from learning what to do when X is presented and what to do when Y is presented. There is a discontinuous jump from the one sort of learning to the other.

In passing, the reader may be interested to know some of the supporting data that would favor the interpretation I am offering.

First, the dog did not show psychotic or neurotic behavior at the beginning of the experiment when he did not know how to discriminate, did not discriminate, and made frequent errors. This did not "break down his discrimination" because he had none, just as at the end the discrimination could not be "broken down" because discrimination was not in fact being asked for.

Second, a naïve dog, offered repeated situations in which some X sometimes means that he is to exhibit behavior A and at other times means that he should exhibit behavior B, will settle down to *guessing*. The naïve dog has not been taught not to guess; that is, he has not been taught that the contexts of life are such that guessing is inappropriate. Such a dog will settle down to reflecting the approximate frequencies of appropriate response. That is, if the stimulus object in 30 percent of cases means A and in 70 percent means B, then the dog will settle down

* This extremely anthropomorphic phrasing is, I claim, not less "objective" than the *ad hoc* abstraction "discrimination."

to exhibiting A in 30 percent of the cases and B in 70 percent. (He will not do what a good gambler would do, namely, exhibit B in all cases.)

Third, if the animals are taken away outside the lab, and if the reinforcements and stimuli are administered from a distance—in the form, for example, of electric shocks carried by long wires lowered from booms (borrowed from Hollywood)—they do not develop symptoms. The shocks, after all, are only of the magnitude of pain that any animal might experience on pushing through a small briar patch; they do not become coercive except in the context of the lab, in which *other* details of the lab (its smell, the experimental stand on which the animal is supported, and so on) become ancillary stimuli that mean to the animal that this is a context in which it *must* continue to be "right." That the animal learns about the nature of laboratory experiment is certainly true, and the same may be said of the graduate student. The experimental subject, whether human or animal, is in the presence of a barrage of *context markers*.

A convenient indicator of logical typing is the reinforcement system to which a given item in our description of behavior will respond. Simple actions apparently respond to reinforcement applied according to the rules of operant conditioning. But *ways of organizing* simple actions, which in our descriptions of behavior we might call "guessing," "discrimination," "play," "exploration," "dependency," "crime," and the like, are of different logical type and do not obey the simple reinforcement rules. The Pavlovian dog could never even be offered affirmative reinforcement for perceiving the change of context because the contrary learning which preceded was so deep and effective.

In the Pavlovian instance, the dog fails to transcend the jump in logical type from "context for discrimination" to "context for guessing."

In contrast, let us consider a case in which an animal achieved a similar jump. At the Oceanic Institute in Hawaii, a female dolphin (*Steno bredanensis*) had been trained to expect the sound of the trainer's whistle to be followed by food and to expect that if she later repeated what she was doing when the whistle blew, she would again hear the whistle and receive food. This animal was being used by the trainers to demonstrate to the public "how we train porpoises."* "When she enters

* "Porpoise" is circus slang for any performing dolphin.

the exhibition tank, I shall watch her and when she does *something* I want her to repeat, I will blow the whistle and she will be fed." She would then repeat her "something" and be again reinforced. Three repetitions of this sequence were enough for the demonstration, and the dolphin was sent offstage to wait for the next performance two hours later. She had learned some simple rules that related her actions, the whistle, the exhibition tank, and the trainer into a pattern, a contextual structure, a set of rules for how to put the information together.

But this pattern was fitted only for a single episode in the exhibition tank. Because the trainers wanted to show again and again how they teach, the dolphin would have to break the simple pattern to deal with the *class* of such episodes. There was a larger *context of contexts* and that would put her in the wrong. At the next performance, the trainer again wanted to demonstrate "operant conditioning," and to do this, she (the trainer) had to pick on a *different* piece of conspicuous behavior. When the dolphin came on stage, she again did her "something," but she got no whistle. The trainer waited for the next piece of conspicuous behavior, perhaps a tail flap, which is a common expression of annoyance. This behavior was then reinforced and repeated.

But the tail flap was, of course, not rewarded in the third performance. Finally, the dolphin learned to deal with the context of contexts by offering a different or *new* piece of conspicuous behavior whenever she came onstage.

All this had happened in the free natural history of the relationship between dolphin and trainer and audience, before I arrived in Hawaii. I saw that what was happening required learning of a higher logical type than usual, and at my suggestion, the sequence was repeated experimentally with a new animal and carefully recorded.* The learning schedule for the experimental training was carefully planned: the animal would experience a series of learning sessions, each lasting from 10 to 20 minutes. The animal would *never* be rewarded for behavior which had been rewarded in the previous session.

Two points from the experimental sequence must be added:

First, it was necessary (in the trainer's judgment) to break the

---

* Described in K. Pryor, R. Haag, and J. O'Reilly, "Deutero-Learning in a Roughtooth Porpoise (*Steno bredanensis*)," U.S. Naval Ordinance Test Station, China Lake, NOTS TP 4270; and further discussed in my *Steps to an Ecology of Mind,* pp. 276–277.

rules of the experiment many times. The experience of being in the wrong was so disturbing to the dolphin that in order to preserve the relationship between her and her trainer (i.e., the context of context of context), it was necessary to give many reinforcements to which the porpoise was not entitled. Unearned fish.

Second, each of the first fourteen sessions was characterized by many futile repetitions of whatever behavior had been reinforced in the immediately preceding session. Seemingly only by accident did the animal provide a piece of different behavior. In the time out between the fourteenth and fifteenth sessions, the dolphin appeared to be much excited; and when she came onstage for the fifteenth session, she put on an elaborate performance that included eight conspicuous pieces of behavior of which four were new and never before observed in this species of animal. From the animal's point of view, there is a jump, a discontinuity, between the logical types.

In all such cases, the step from one logical type to the next higher is a step from information about an event to information about a class of events or from considering the class to considering the class of classes. Notably, in the case of the dolphin, it was impossible for her to learn from a single experience, whether of success or failure, that the context was one for exhibiting a *new* behavior. The lesson about context could only have been learned from comparative information about a sample of contexts differing among themselves, in which her behavior and the outcome differed from instance to instance. Within such a varied class, a regularity became perceptible, and the apparent contradiction could be transcended. The case of the dog would have involved a similar step, but the dog did not have a chance to learn that this was a situation for guesswork.

Much can be learned from a single instance, but not certain things about the nature of the larger sample, the class, of such trials or experiences. This is fundamental for logical typing, whether at the level of Bertrand Russell's abstractions or at the level of animal learning in a real world.

That these are not phenomena relevant only to the laboratory and animal learning experiments may be driven home by calling attention to some human confusions of thought. A number of concepts are freely bandied about by layman and expert alike with an implicit error in their logical typing. For example, there is "exploration." It seems to puzzle psychologists that the exploring tendencies of a rat cannot be simply extinguished by having the rat encounter boxes containing small electric shocks. From such experiences, the rat will not learn not to put his nose into boxes; he will only learn not to put his nose into the particular boxes that contained electric shocks when he investigated them. In other words, we are here up against a contrast between learning about the particular and learning about the general.

A little empathy will show that from the rat's point of view, it is not desirable that he learn the general lesson. His experience of a shock upon putting his nose into a box indicates to him that he did *well* to put his nose into that box in order to gain the information that it contained a shock. In fact, the "purpose" of exploration is, not to discover whether exploration is a good thing, but to discover information about the explored. The larger case is of a totally different nature from that of the particular.

It is interesting to consider the nature of such a concept as "crime." We act as if crime could be extinguished by punishing parts of what we regard as criminal actions, as if "crime" were the name of a sort of action or of part of a sort of action. More correctly "crime," like "exploration," is the name of a way of organizing actions. It is therefore unlikely that punishing the act will extinguish the crime. In several thousand years, the so-called science of criminology has not escaped from a simple blunder in logical typing.

Be that as it may, there is a very profound difference between a serious attempt to change the characterological state of an organism and trying to change that organism's particular actions. The latter is relatively easy; the former, profoundly difficult. Paradigmatic change is as difficult as—indeed is of the same nature as—change in epistemology. (For an elaborate study of what seems to be necessary to make characterological changes in human criminals, the reader is referred to a recent book, *Sane Asylum,* by Charles Hampden-Turner.*) It would seem to be almost a first requirement of such deep training that the particular act for which the convict was being punished when in jail should *not* be the main focus of the training.

---

* Charles Hampden-Turner, *Sane Asylum* (San Francisco: San Francisco Book Co., 1976).

A third concept of the class which is commonly misunderstood by wrong attribution of logical typing is "play." The given acts that constitute play in a given sequence may, of course, occur in the same persons or animals in other sorts of sequence. What is characteristic of "play" is that this is a name for contexts in which the constituent acts have a different sort of relevance and organization from that which they would have had in non-play. It may even be that the essence of play lies in a partial denial of the meanings that the actions would have had in other situations. It was from a recognition that mammals recognize play that I moved forward twenty years ago to a recognition that animals (in that case, river otters) classify their types of interchange and therefore are subject to the sorts of pathology generated in the Pavlovian dog who is punished for a failure to recognize a change of context or the criminal who is made to suffer for particular acts when he or she should be suffering for particular ways of organizing action. From observation of play in river otters, I went on to study similar classifications of behavior in human beings, finally arriving at the notion that certain symptoms of human pathology called *schizophrenia* were, in fact, also the outcome of maltreatments of logical typing, which we called *double binds*.

In this section, I have approached the matter of hierarchy in mental phenomena from the aspect of coding. But hierarchy could equally well have been demonstrated from criterion 4, which deals with circular chains of determination. The relationship between the characteristics of a component and the characteristics of the system as a whole as it circles back on itself, is equally a matter of hierarchical organization.

I want to suggest here that the history of civilization's long flirtation with the notion of circular cause would seem to be shaped by the partial fascination and partial terror associated with the matter of logical typing. It was noted in Chapter 2 (section 13) that logic is a poor model of cause and effect. I suggest that it is the attempt to deal with life in logical terms and the compulsive nature of that attempt which produce in us the propensity for terror when it is even hinted that such a logical approach might break down.

In Chapter 2, I argued that the very simple buzzer circuit, if spelled out onto a logical map or model, presents contradiction: If the buzzer circuit is closed, then the armature is attracted by the electromagnet. If the armature moves, attracted by the electromagnet, the

attraction ceases, and the armature is then not attracted. This cycle of *if*
. . . *then* relations in the world of cause is disruptive of any cycle of *if*
. . . *then* relations in the world of logic unless time is introduced into
logic. The disruption is formally similar to the paradox of Epimenides.

We humans seem to wish that our logic were absolute. We seem
to act on the assumption that it is so and then panic when the slightest
overtone that it is not so, or might not be so, is presented.

It is as if the tight coherence of the logical brain, even in persons
who notoriously think with a great deal of muddleheadedness, must still
be sacrosanct. When it is shown to be not so coherent, the individuals or
cultures dash precipitately, like Gadarene swine, into complexities of su-
pernaturalism. In order to escape the million metaphoric deaths depicted
in a universe of *circles* of causation, we are eager to deny the simple reality of ordinary dying and to build fantasies of an afterworld and even of
reincarnation.

In truth, a breach in the apparent coherence of our mental logical process would seem to be a sort of death. I encountered this deep notion over and over again in my dealings with schizophrenics, and the notion may be said to be basic to the double bind theory that I and my
colleagues at Palo Alto proposed some twenty years ago.* I am proposing here that the hint of death is present in every biological circuit
whatsoever.

To conclude this chapter, I shall mention some of the potential-
ities of minds that exhibit these six criteria. First of all, there are two
characteristics of mind that may be mentioned together, both of which
are made possible by the criteria I have cited. These two closely related
characteristics are autonomy and death.

Autonomy—literally *control of the self,* from the Greek *autos* (self)
and *nomos* (a law)—is provided by the recursive structure of the system.
Whether or not a simple machine with a governor can control or be controlled by itself may be disputed, but imagine more loops of information
and effect added on top of the simple circuit. What will be the content

* I was lucky enough at that time to obtain a copy of John Perceval's account of his psychosis in the
1830s. This book is now available as *Perceval's Narrative* and shows how the schizophrenic's world is
totally structured in double bind terms. (John Perceval, *Perceval's Narrative: A Patient's Account of His
Psychosis, 1830–32,* Gregory Bateson, ed. Stanford, Calif.: Stanford University Press, 1961.)

of the signal material carried by these loops? The answer, of course, is that these loops will carry messages *about* the behavior of the whole system. In a sense, the original simple circuit already contained such information ("It's going too fast"; "it's going too slow"), but the next level will carry such information as "the correction of 'it's going too fast' is not fast enough," or "the correction of 'it's going too fast' is excessive." That is, the messages become messages about the previous lower level. From this to autonomy is a very short step.

With regard to death, the possibility for death follows first from criterion 1, that the entity be made of multiple parts. In death, these parts are disassembled or randomized. But it arises also from criterion 4. Death is the breaking up of the circuits and, with that, the destruction of autonomy.

In addition to these two very profound characteristics, the sort of system that I call *mind* is capable of purpose and choice by way of its self-corrective possibilities. It is capable of either steady state or runaway or some mixture of these. It is influenced by "maps," never by territory, and is therefore limited by the generalization that its receipt of information will never *prove* anything about the world or about itself. As I stated in Chapter 2, science never proves anything.

Beyond this, the system will learn and remember, it will build up negentropy, and it will do so by the playing of stochastic games called *empiricism* or *trial and error*. It will store energy. It will inevitably be characterized by the fact that all messages are of some logical type or other, and so it will be dogged by the possibilities of error in logical typing. Finally, the system will be capable of uniting with other similar systems to make still larger wholes.

In conclusion, two questions may be raised: Will the system be capable of some sort of aesthetic preference? Will the system be capable of consciousness?

With regard to aesthetic preference, it seems to me that the answer could be affirmative. It is conceivable that such systems would be able to recognize characteristics similar to their own in other systems they might encounter. It is conceivable that we may take the six criteria as criteria of life and may guess that any entity exhibiting these characteristics will set a value (*plus* or *minus*) on other systems exhibiting the outward and visible signs of similar characteristics. Is our reason for admiring a daisy the fact that it shows—in its form, in its growth, in its coloring, and in its death—the symptoms of being alive? Our appreciation for it is to that extent an appreciation of its similarity to ourselves.

With regard to consciousness, the matter is more obscure. In this book, nothing has been said about consciousness except to note that in the business of perception, the processes of perception are not conscious but that its products may be conscious. When *consciousness* is used in this sense, it would appear that the phenomenon is somehow related to the business of logical types to which we have given a good deal of attention. However, I do not know of any material really connecting the phenomena of consciousness to more primitive or simpler phenomena and have not attempted to do so in the present work.

V

MULTIPLE
VERSIONS
OF
RELATIONSHIP

If they be two, they two are so
   As stiffe twin compasses are two;
Thy soule, the fixt foot, makes no show
   To move, but doth if th' other doe.

And though it in the center sit,
   Yet when the other far doth rome,
It leanes, and hearkens after it,
   And growes erect, as that comes home.

Such wilt thou be to me, who must
   Like th' other foot, obliquely runne.
Thy firmnes drawes my circle just,
   And makes me end where I begunne.
   —JOHN DONNE, *"A Valediction: Forbidding Mourning"*

In Chapter 3, I considered the working together of two eyes to give binocular vision. From the combined vision of the two organs, you get a species of information that you could get from a single eye only by using special sorts of collateral knowledge (e.g., about the overlapping of things in the visual field); you get, in fact, depth perception. This is information about a different dimension (as the physicist would call it) or information of a different logical type (as I would call it).

In this chapter, in addition to talking about double description, I want to examine the subject of boundaries. What limits the units, what limits "things," and above all, what, if anything, *limits the self?*

Is there a line or sort of bag of which we can say that "inside" that line or interface is "me" and "outside" is the environment or some other person? By what right do we make these distinctions?

It is clear (though usually ignored) that the language of any answer to that question is *not,* in the end, a language of space or time. "Inside" and "outside" are not appropriate metaphors for inclusion and exclusion when we are speaking of the self.

The mind contains no things, no pigs, no people, no midwife toads, or what have you, only ideas (i.e., news of difference), information about "things" in quotes, always in quotes. Similarly, the mind contains no time and no space, only ideas of "time" and "space." It follows that the boundaries of the individual, if real at all, will be, not spatial boundaries, but something more like the sacks that represent *sets* in set theoretical diagrams or the bubbles that come out of the mouths of the characters in comic strips.

My daughter, now aged ten, had her birthday last week. The tenth birthday is an important one because it represents a breakthrough into two-digit numbers. She remarked, half serious and half in jest, that she did not "feel any different."

The boundary between the ninth year and the tenth year was not *real* in the sense of being or representing a change in feeling. But one could perhaps make Venn diagrams or bubbles to *classify* propositions about various ages.

In addition, I want to focus on that genus of *receipt of information* (or call it *learning*) which is learning about the "self" in a way that may result in some "change" in the "self." Especially, I will look at changes in the boundaries of the self, perhaps at the discovery that there are boundaries or perhaps no center. And so on.

How do we learn those learnings or wisdoms (or follies) by which "we ourselves"—our ideas about self—seem to be changed?

I began to think about such matters a long time ago, and here are two notions that I developed before World War II, when I was working out what I called the "dynamics" or "mechanics" of Iatmul culture on the Sepik River in New Guinea.

One notion was that the unit of *interaction* and the unit of *characterological learning* (not just acquiring the so-called "response" when the buzzer sounds, but the *becoming ready for such automatisms*) are the same.

Learning the contexts of life is a matter that has to be discussed, not internally, but as a matter of the external relationship between two creatures. And *relationship is always a product of double description.*

It is correct (and a great improvement) to begin to think of the two parties to the interaction as two eyes, each giving a monocular view of what goes on and, together, giving a binocular view in depth. This double view *is* the relationship.

Relationship is not internal to the single person. It is nonsense to talk about "dependency" or "aggressiveness" or "pride," and so on. All such words have their roots in what happens between persons, not in some something-or-other inside a person.

No doubt there is a learning in the more particular sense. There are changes in A and changes in B which correspond to the dependency-succorance of the relationship. But the relationship comes first; it *precedes*.

Only if you hold on tight to the primacy and priority of relationship can you avoid dormitive explanations. The opium does not contain a dormitive principle, and the man does not contain an aggressive instinct.

The New Guinea material and much that has come later, taught me that I will get nowhere by explaining prideful behavior, for example, by referring to an individual's "pride." Nor can you explain aggression by referring to instinctive (or even learned) "aggressiveness."* Such an explanation, which shifts attention from the interpersonal field to a factitious inner tendency, principle, instinct, or whatnot, is, I suggest, very great nonsense which only hides the real questions.

If you want to talk about, say, "pride," you must talk about two persons or two groups and what happens between them. A is admired by B; B's admiration is conditional and may turn to contempt. And so on. You can then define a particular species of pride by reference to a particular pattern of interaction.

The same is true of "dependency," "courage," "passive-aggressive behavior," "fatalism," and the like. *All* characterological adjectives are to be reduced or expanded to derive their definitions from patterns of interchange, i.e., from combinations of double description.

As binocular vision gives the possibility of a new order of information (about depth), so the understanding (conscious and unconscious) of behavior through relationship gives a new *logical type* of learning. (In

* Note, in passing, how easy is the descent from sociobiology to paranoia and, perhaps, how easy is the descent from violent repudiation of sociobiology to paranoia—alas.

Steps to an Ecology of Mind, I have called this Learning II, or *deutero-learning.*)

The whole matter is a little difficult to grasp because we have been taught to think of learning as a two-unit affair: The teacher "taught," and the student (or the experimental animal) "learned." But that lineal model became obsolete when we learned about cybernetic circuits of interaction. The minimum unit of interaction contains three components. (In this, the old experimenters were right, in spite of their blindness to differences in logical levels.)

Call the three components *stimulus, response,* and *reinforcement.* Of these three, the second is the reinforcement of the first, and the third is reinforcement of the second. *Response* by *learner* reinforces the *stimulus* provided by *teacher.* And so on.

Pride is conditional admiration provided by spectator, *plus* response by performer, *plus* more admiration, *plus* acceptance of admiration. . . . (Cut the sequence where you will!) Of course, there are hundreds of ways in which the components of the contexts of learning may be interlinked, and, correspondingly, hundreds of characterological "traits," of which hundreds the experimenters have looked at about half a dozen—strange.

I am saying that there is a learning of context, a learning that is different from what the experimenters see. And that this learning of context springs out of a species of double description which goes with relationship and interaction. Moreover, like all themes of contextual learning, these themes of relationship are self-validating. Pride feeds on admiration. But because the admiration is conditional—and the proud man fears the contempt of the other—it follows that there is nothing which the other can do to diminish the pride. If he shows contempt, he equally reinforces the pride.

Similarly, we can expect self-validation in other examples of the same logical typing. Exploration, play, crime, and the Type A behavior of the psychosomatic studies of hypertension are equally difficult to extinguish. Of course, all these are not categories of behavior; they are *categories of contextual organization of behavior.*

In summary, this chapter adds important generalizations. We now see that the mechanics of relationship are a special case of double description and that the unit of behavioral sequence contains at least three components, maybe many more.

1. "KNOW THYSELF"

The old Greek advice "know thyself" may carry many levels of mystic insight, but in addition to these aspects of the matter, there is a very simple, universal and, indeed, pragmatic aspect. It is surely so that all outside knowledge whatsoever must derive in part from what is called *self-knowledge*.

The Buddhists claim that the self is a sort of fiction. If so, our task will be to identify the species of fiction. But for the moment, I shall accept the "self" as a heuristic concept, a ladder useful in climbing but perhaps to be thrown away or left behind at a later stage.

I reach out with my hand in the dark, and it touches the electric light switch. "I have found it. That's where *it* is"; and *"I* can now turn *it* on."

But I did not need to know the position of the switch or the position of my hand to be able to turn the light on. The mere sensory report of contact between hand and switch would have been enough. I could have been in total error in my "that's where it is," and still, with my hand on the switch, I could turn it on.

The question is: *Where is my hand?* This item of self-knowledge has a very special and peculiar relation to the business of searching for the switch or *knowing* where the switch is.

Under hypnosis, for example, I could have believed that my hand was above my head when, in fact, it was stretched horizontally forward. In such a case, I would have located the switch up there, above my head. I might even have taken my success in turning on the light as a verification of my discovery that the switch was "above my head."

We *project* our opinions of self onto the outer world, and often we can be wrong about the self and still move and act and interact with our friends successfully but in terms of false opinions.

What, then, is this "self"? What, in the context of the present chapter, is added to information by obeying the old advice "know thyself"?

Let me start again. Suppose that I "know" that my hand is above my head and that I "know" the light switch is at shoulder height. Let us suppose that I am right about the switch but wrong about my hand. In the search for the switch, I shall never put my hand where the switch is. It would be better if I did not "know" the position of the switch. I would then perhaps find it by some random movement of trial and error.

What, then, are the rules for self-knowledge? Under what circumstances is it (pragmatically) better to have no such knowledge than to have erroneous opinions? Under what circumstances is self-knowledge pragmatically necessary? Most people seem to live without any answers to questions of this sort. Indeed, they seem to live without even asking such questions.

Let us approach the whole matter with less epistemological arrogance. Does a dog have self-knowledge? Is it possible that a dog with *no* self-knowledge can chase a rabbit? Is the whole mass of injunctions that tell us to know ourselves just a tangle of monstrous illusions built up to compensate for the paradoxes of consciousness?

If we throw away the notion that the dog is one creature and the rabbit another and consider the whole rabbit-dog as a single system, we can now ask: What redundancies must exist in this system so that this part of the system will be able to chase that part? And, perhaps, be unable to *not* chase it?

The answer now appears to be quite different: The only information (i.e., redundancy) that is necessary in these cases is relational. Did the rabbit, by running, *tell* the dog to chase it? In the matter of turning on the light, when the hand ("my" hand?) touched the switch, the necessary information about *relationship* between hand and switch was created; and turning on the switch became possible without collateral information about me, my hand, or the switch.

The dog can invite to a game of "chase me." He goes down with his chin and throat to the ground and reaches forward, with his front legs, from elbows to pads, pressed against the ground. His eyes look up, moving in their sockets without any movement of the head. The hind legs are bent under the body ready to spring forward. This posture is familiar to anybody who has ever played with a dog. The existence of such a signal proves the dog able to communicate at, at least, two Russellian levels or logical types.

Here, however, I am concerned only with those aspects of play which exemplify the rule that *two descriptions are better than one.*

The game and the creation of the game must be seen as a single phenomenon, and indeed, it is subjectively plausible to say that the sequence is really playable only so long as it retains some elements of the creative and unexpected. If the sequence is totally known, it is *ritual,* although perhaps still character forming.* It is rather simple to see a first level of discovery by human player, A, who has a finite number of alternative actions. These are evolutionary sequences, with natural selection of, not items, but *patterns of items* of action. A will try various actions on B and find that B will only accept certain contexts. That is to say, A must either precede certain actions with certain others or place certain of his own actions into time frames (sequences of interaction) that are preferred by B. A "proposes"; B "disposes."

A superficially miraculous phenomenon is the invention of play between members of contrasting mammalian species. I have watched this process in interaction between our keeshond and our tame gibbon, and it was quite clear that the dog responded in her normal way to an unexpected tweak of the fur. The gibbon would come suddenly out of the rafters of the porch roof and lightly attack. The dog would give chase, the gibbon would run away, and the whole system would move from the porch to our bedroom, which had a ceiling instead of exposed rafters and beams. Confined to the floor, the retreating gibbon would turn on the dog, who would retreat, running out onto the porch. The gibbon would then go up into the roof, and the whole sequence would start over again, to be repeated many times and evidently enjoyed by both players.

Discovering or inventing games with a dolphin in the water is a very similar experience. I had decided to give the elderly female *Tursiops* no clues about how to deal with me other than the "stimulus" of my presence in the water. So I sat, with arms folded, on the steps leading down into the water. The dolphin came over and stationed herself alongside me, about one or two inches away from contact with my side. From time to time, there would be accidental physical contact between us due to movements of the water. These contacts were seemingly of no interest

* If we define play as the establishment and exploration of relationship, then greeting and ritual are the affirmation of relationship. But obviously mixtures of affirmation and exploration are common.

to the animal. After perhaps two minutes, she moved away and slowly swam around me; and a few moments later, I felt something pushing in under my right arm. This was the dolphin's beak, and I was confronted with a problem: how to give the animal *no clues* about how to deal with me. My planned strategy was impossible.

I relaxed my right arm and let her push her beak under it. In seconds, I had a whole dolphin under my arm. She then bent around in front of me to a position in which she was sitting in my lap. From this position, we went on to a few minutes of swimming and playing together.

Next day, I followed the same sequence but did not wait out the period of minutes while she was alongside. I stroked her back with my hand. She immediately corrected me, swimming a short distance away and then circling me and giving me a flick with the leading edge of her tail fluke, no doubt an act that seemed to her to be gentle. After that, she went to the far end of the tank and stayed there.

Again, these are evolutionary sequences, and it is important to see clearly just *what* is evolved. To describe the cross-species play of dog-and-gibbon or man-and-dolphin as an evolution of items of behavior would be incorrect because no new items of behavior are generated. Indeed, for each creature in turn, there is no evolution of new contexts of action. The dog is still unchanged dog; the gibbon is still gibbon; the dolphin, dolphin; the man, man. Each retains its own "character"—its own organization of the perceived universe—and yet, clearly something has happened. Patterns of interaction have been generated or discovered, and these patterns have, at least briefly, endured. In other words, there has been a natural selection of patterns of interaction. Certain patterns survived longer than others.

There has been an evolution of *fitting together.* With minimum change in dog or gibbon, the system dog-gibbon has become simpler—more internally integrated and consistent.

There is thus a larger entity, call it *A plus B,* and that larger entity, in play, is achieving a process for which I suggest that the correct name is *practice.* This is a learning process in which the system A plus B receives no new information from outside, only from *within the system.* The interaction makes information about parts of A available to parts of B and *vice versa.* There has been a change in boundaries.

Let us place these data in a wider theoretical frame. Let us do a little *abduction,* seeking other cases which will be anologous to play in the sense of belonging under the same rule.

Notice that *play,* as a label, does not limit or define the acts that make up play. *Play* is applicable only to certain broad premises of the interchange. In ordinary parlance, "play" is not the name of an act or action; it is the name of a *frame* for action. We may expect, then, that play is not subject to the regular rules of reinforcement. Indeed, anybody who has tried to stop some children playing knows how it feels when his efforts simply get included in the shape of the game.

So to find other cases under the same rule (or chunk of theory), we look for integrations of behavior which a) do not define the actions which are their content; and b) do not obey the ordinary reinforcement rules.

Two cases come immediately to mind: "exploration" and "crime." Others worth thinking about are "Type A behavior" (which the psychosomatic doctors regard as partly etiological for essential hypertension), "paranoia," "schizophrenia," and so on.

Let us examine "exploration" to see wherein it is a context for, or a product of, some sort of double description.

First, exploration (and crime and play and all the other words of this class) is a primary description, verbal or nonverbal, of the self: *"I* explore." But *what* is explored is not merely "my outside world," or "the outside world as *I* live it."

Second, exploration is self-validating, whether the outcome is pleasant or unpleasant for the explorer. If you try to teach a rat to not-explore by having him poke his nose into boxes containing electric shock, he will, as we saw in the last chapter, go on doing this, presumably needing to know which boxes are safe and which unsafe. In this sense, exploration is always a success.

Thus, exploration is not only self-validating; it also seems in human beings to be addictive. I once knew a great mountain climber, Geoffrey Young, who climbed the north face of the Matterhorn with only one leg. (The other had been amputated in World War I.) And I knew a long-distance runner, Leigh Mallory, whose bones are somewhere within 200 feet of the top of Mount Everest. These climbers give us a hint about exploration. Geoffrey Young used to say that the *not-listening* to the weak and self-pitying complaints and pains of the body was among the main disciplines of the climber—even, I think, among the satisfactions of climbing. The victory over self.

Such changing of "self" is commonly described as a "victory," and such lineal words as "discipline," and "self-control" are used. Of course these are mere supernaturalisms—and probably a little toxic at that. What happens is much more like an incorporation or marriage of ideas about the world with ideas about self.

This brings up another example, traditionally familiar to anthropologists: totemism.

2. TOTEMISM

For many peoples, their thinking about the social system of which they are the parts is shaped (literally in-formed) by an analogy between that system of which they *are* the parts and the larger ecological and biological system in which the animals and plants and the people are all parts. The analogy is partly exact and partly fanciful and partly made real—validated—by actions that the fantasy dictates. The fantasy then becomes morphogenetic; that is, it becomes a determinant of the shape of the society.

This analogy between the social system and the natural world is the religion that anthropologists call *totemism*. As analogy, it is both more appropriate and more healthy than the analogy, familiar to us, which would liken people and society to nineteenth-century machines.

In its late and partly secular form, totemism is familiar to the occidental world as the premise of heraldry. Families or patrilineal lines claim ancient dignity by depicting animals on their heraldic shields or totem poles, which thus become genealogical diagrams by the combining of the beasts of different ancestral lines. Such representations of family status in a mythological hierarchy often aggrandize self or own descent at the expense of other family lines. As this more prideful component of totemism increases, the larger view of relationship to the natural world is likely to be forgotten or reduced to a mere pun. My own family has a crest, granted in the eighteenth century. It is, of course, a bat's wing. Similarly, my father's mother's Lowland Scots family, whose name was Aikin, had an oak tree emblazoned on their silverware. In their dialect, it is proverbial that "from little aikins [i.e., acorns] big aiks grow." And so on.

What seems to happen in such conventional secularization is a shift of attention away from the relationship to focus *one end,* on the objects or persons who were related. This is a common pathway leading to vulgarized epistemology and to a loss of that insight or enlightenment which was gained by setting the view of nature beside the view of family.

However, there are still a few practicing totemites, even in the ranks of professional biology. To watch Professor Konrad Lorenz teach a class is to discover what the Aurignacian cavemen were doing when they painted those living, moving reindeer and mammoths on the sides and ceilings of their caves. Lorenz's posture and expressive movement, his kinesics, change from moment to moment according to the nature of the animal he is talking about. At one moment, he is a goose; a few minutes later, a cichlid fish. And so on. He will go to the blackboard and quickly draw the creature, perhaps a dog, alive and hesitating between attack and retreat. Then a moment's work with eraser and chalk, a change in the back of the neck and the angle of the tail, and the dog is now clearly going to attack.

He gave a series of lectures in Hawaii and devoted the last of these to problems of the philosophy of science. When he spoke of the Einsteinian universe, his body seemed to twist and contort a little in empathy with that abstraction.

And mysteriously, like the Aurignacians, he is unable to draw a human figure. His attempts and theirs result only in stickmen. What totemism teaches about the self is profoundly nonvisual.

Lorenz's empathy with animals gives him an almost unfair advantage over other zoologists. He can, and surely does, read much from a (conscious or unconscious) comparison of what he sees the animal do with what it feels like to do the same. (Many psychiatrists use the same trick to discover the thoughts and feelings of their patients.) Two diverse descriptions are always better than one.

Today, we can stand back from the double description that is the native totemism of aboriginal Australia and from the totemism of European heraldry and look at the *process* of degeneration. We can see how ego displaced enlightenment, how the family animals became crests and banners, and how the relations between the animal prototypes in nature got forgotten.

(Today, we pump a little natural history into children along with a little "art" so that they will forget their animal and ecological nature and the aesthetics of being alive and will grow up to be good businessmen.)

There is, by the way, another pathway of degeneracy that becomes visible in the *comparative* survey we are discussing. This is the Aesop-ation of natural history. In this process, it is not pride and ego but *entertainment* that replaces religion. The natural history is no longer even a pretense of looking at real creatures; it becomes a cluster of stories, more or less sardonic, more or less moral, more or less amusing. The holistic view that I am calling *religion* splits to give either weapons to ego or toys to fancy.

3. ABDUCTION

We are so accustomed to the universe in which we live and to our puny methods of thinking about it that we can hardly see that it is, for example, surprising that abduction is possible, that it is possible to describe some event or thing (e.g., a man shaving in a mirror) and then to look around the world for other cases to fit the same rules that we devised for our description. We can look at the anatomy of a frog and then look around to find other instances of the same abstract relations recurring in other creatures, including, in this case, ourselves.

This lateral extension of abstract components of description is called *abduction,* and I hope the reader may see it with a fresh eye. The very possibility of abduction is a little uncanny, and the phenomenon is enormously more widespread than he or she might, at first thought, have supposed.

Metaphor, dream, parable, allegory, the whole of art, the whole of science, the whole of religion, the whole of poetry, totemism (as already mentioned), the organization of facts in comparative anatomy—all these are instances or aggregates of instances of abduction, within the human mental sphere.

But obviously, the possibility of abduction extends to the very roots also of physical science, Newton's analysis of the solar system and the periodic table of the elements being historic examples.

Conversely, all thought would be totally impossible in a universe in which abduction was not expectable.

Here I am concerned only with that aspect of the universal fact of abduction which is relevant to the order of change that is the subject of this chapter. I am concerned with changes in basic epistemology, character, self, and so on. Any change in our epistemology will involve shifting our whole system of abductions. We must pass through the threat of that chaos where thought becomes impossible.

Every abduction may be seen as a double or multiple description of some object or event or sequence. If I examine the social organization of an Australian tribe and the sketch of natural relations upon which the totemism is based, I can see these two bodies of knowledge as related abductively, as both falling under the same rules. In each case, it is assumed that certain formal characteristics of one component will be mirrored in the other.

This repetition has certain very effective implications. It carries injunctions, for the people concerned. Their ideas about nature, however fantastic, are supported by their social system; conversely, the social system is supported by their ideas of nature. It thus becomes very difficult for the people, so doubly guided, to change their view either of nature or of the social system. For the benefits of stability, they pay the price of rigidity, living, as all human beings must, in an enormously complex network of mutually supporting presuppositions. The converse of this statement is that change will require various sorts of relaxation or contradiction within the system of presuppositions.

What seems to be the case is that there are, in nature and correspondingly reflected in our processes of thought, great regions within which abductive systems obtain. For example, the anatomy and physiology of the body can be considered as one vast abductive system with its own coherence within itself at any given time. Similarly, the environment within which the creature lives is another such internally coherent abductive system, although this system is not immediately coherent with that of the organism.

For change to occur, a double requirement is imposed on the

new thing. It must fit the organism's internal demands for coherence, and it must fit the external requirements of environment.

It thus comes about that what I have called *double description* becomes double requirement or double specification. The possibilities for change are twice fractionated. If the creature is to endure, change must always occur in ways that are doubly defined. Broadly, the internal requirements of the body will be conservative. Survival of the body requires that not-too-great disruption shall occur. In contrast, the changing environment may require change in the organism and a sacrifice of conservatism.

In Chapter 6, we shall consider the resulting contrast between homology, which is the result of phylogenetic conservatism, and adaptation, which is the reward of change.

VI

THE
GREAT
STOCHASTIC
PROCESSES

The expression often used by Mr. Herbert Spencer of the Survival of the Fittest is more accurate, and is sometimes equally convenient.
—CHARLES DARWIN, *On the Origin of Species,* FIFTH EDITION.

Into this universe, and why not knowing
Nor whence, like Water willy-nilly flowing:
And out of it, as Wind along the Waste,
I know not whither, willy-nilly blowing.
—EDWARD FITZGERALD, *The Rubaiyat of Omar Khayyam*

It is a general assumption of this book that both genetic change and the process called *learning* (including the somatic changes induced by habit and environment) are stochastic processes. In each case there is, I believe, a stream of events that is random in certain aspects and in each case there is a nonrandom selective process which causes certain of the random components to "survive" longer than others. Without the random, there can be no new thing.

I assume that in evolution the production of mutant forms is either random within whatever set of alternatives the status quo ante will permit or that, if mutation be ordered, the criteria of that ordering are irrelevant to the stresses of the organism. In accordance with orthodox molecular genetic theory, I assume that the protoplasmic environment of the DNA cannot direct changes in DNA which would be relevant to fitting the organism to the environment or reducing internal stress. Many factors—both physical and chemical—can alter the frequency of mutation, but I assume that the mutations so generated are not geared to the particular stresses which the parent generation was under at the time when the mutation was brought about. I shall even assume that mutations produced by a mutagen are irrelevant to the physiological stress generated within the cell by the mutagen itself.

Beyond that, I shall assume, as is now orthodox, that mutations, so randomly generated, are stored in the mixed gene pool of the population and that natural selection will work to eliminate those alternatives which are unfavorable from the point of view of *something like* survival and that this elimination will, on the whole, favor those alternatives which are harmless or beneficial.

On the side of the individual, I similarly assume that the mental processes generate a large number of alternatives and that there is a selection among these determined by *something like* reinforcement.

Both for mutations and for learning, it is always necessary to remember the potential pathologies of logical typing. What has survival value for the individual may be lethal for the population or for the society. What is good for a short time (the symptomatic cure) may be addictive or lethal over long time.

It was Alfred Russel Wallace who remarked in 1866 that the principle of natural selection is like that of the steam engine with a governor. I shall assume that this is indeed so and that both the process of individual learning and the process of population shift under natural selection can exhibit the pathologies of all cybernetic circuits: excessive oscillation and runaway.

In sum, I shall assume that evolutionary change and somatic change (including learning and thought) are fundamentally similar, that both are stochastic in nature, although surely the ideas (injunctions, descriptive propositions, and so on) on which each process works are of totally different logical typing from the typing of ideas in the other process.

It is this tangle of logical typing that has led to so much confusion, controversy, and even nonsense about such matters as the "inheritance of acquired characteristics" and the legitimacy of invoking "mind" as an explanatory principle.

The whole matter has had a curious history. It was once intolerable to many people to suggest that evolution could have a random component. This was supposedly contrary to all that was known about adaptation and design and contrary to any belief in a creator with mental characteristics. Samuel Butler's criticism of *The Origin of Species* was essentially to accuse Darwin of excluding mind from among the relevant explanatory principles. Butler wanted to imagine a nonrandom mind at work somewhere in the system and therefore preferred the theories of Lamarck to those of Darwin.*

It turns out, however, that such critics were precisely wrong in their choice of the correction they would apply to Darwinian theory. Today, we see thought and learning (and perhaps somatic change) as stochastic. We would correct the nineteenth-century thinkers, not by adding a nonstochastic mind to the evolutionary process, but by proposing that thought and evolution are alike in a shared stochasticism. Both are mental processes in terms of the criteria offered in Chapter 4.

We face, then, two great stochastic systems that are partly in interaction and partly isolated from each other. One system is within the individual and is called *learning;* the other is immanent in heredity and in populations and is called *evolution.* One is a matter of the single lifetime; the other is a matter of multiple generations of many individuals.

The task of this chapter is to show how these two stochastic systems, working at different levels of logical typing, fit together into a single ongoing biosphere that could not endure if either somatic or genetic change were fundamentally different from what it is.

The *unity* of the combined system is *necessary.*

1. THE LAMARCKIAN ERRORS

A very large part of what can be said about the interlocking of evolution and somatic change is deductive. At the levels of theory that we confront here, there are no observational data, and experimentation

---

* Strangely, even in Butler's *Evolution, Old and New,* there is very little evidence that Butler had much empathy for the delicate thinking of Lamarck.

has not yet begun. But this is not surprising. There was, after all, almost no field evidence for natural selection until Kettlewell studied the pale and melanic varieties of pepper moth (*Biston betularia*) in the 1930s.

In any case, the arguments against the hypothesis that acquired characteristics are inherited are instructive and will serve to illustrate several aspects of the tangled relationship between the two great stochastic processes. There are three such arguments, of which only the third is convincing:

a. The first argument is that the hypothesis is to be discarded for lack of empirical support. But experimentation in this field is incredibly difficult and the critics ruthless, so the lack of evidence is not surprising. It is not clear that if Lamarckian inheritance occurred either in the field or even in the laboratory, it would be possible to recognize it.

b. The second and until recently the most cogent criticism was August Weissmann's assertion in the 1890s that there is *no communication between soma and germ plasm.* Weissmann was an extraordinarily gifted German embryologist who, becoming nearly blind while still a young man, devoted himself to theory. He noted that for many organisms there was a continuity of what he called "germ plasm," i.e., of the protoplasmic line from generation to generation, and that in each generation the phenotypic body or soma could be considered as branching off from this germ plasm. From this insight he argued that there could be no backward communication from the somatic branch to the main stem which was the germ plasm.

Exercise of the right biceps will certainly strengthen that muscle in an individual, but there is no known way in which news of that somatic change could be carried to the sex cells of that individual. This criticism, like the first, depends on argument from the fact of absence of evidence—an unsure stone on which to step—and most biologists after Weissmann have tended to make the argument *deductive* by assuming that there is *no imaginable* way in which communication could occur between biceps and future gamete.

But that assumption does not look so safe today as it did twenty years ago. If RNA can carry imprints of portions of DNA to other parts of the cell and possibly to other parts of the body, then it is *imaginable* that imprints of chemical changes in the biceps could be carried to the germ plasm.

c. The final and, for me, the only convincing criticism is a reductio ad absurdum, an assertion that if Lamarckian inheritance were the rule or even at all common, the whole system of interlocking stochastic processes would come to a halt.

I offer this criticism here not only in an attempt (probably futile) to kill a never-quite-dead horse but also to illustrate the relations between the two stochastic processes. Imagine the following dialogue:

BIOLOGIST: What exactly is claimed by Lamarckian theory? What do you mean by *"the inheritance of acquired characteristics"?*

LAMARCKIAN: That a change in the body induced by environment will be passed on to the offspring.

BIOLOGIST: Wait a minute, a *"change"* is to be passed on? What exactly is to be passed from parent to offspring? A "change" is some sort of abstraction, I suppose.

LAMARCKIAN: An effect of environment, for example, the nuptial pads of the male midwife toad.*

BIOLOGIST: I still don't understand. You surely do not mean that the environment made the nuptial pads.

LAMARCKIAN: No, of course not. The toad made them.

BIOLOGIST: Ah, so the toad knew in some sense or had the "potentiality" for growing nuptial pads?

LAMARCKIAN: Something like that, yes. The toad could make nuptial pads when forced to breed in water.

BIOLOGIST: Ah, he could adapt himself. Is that right? If he bred on land, in the way normal to his species of toad, he made no nuptial pads. If in water, then he made pads just like all the other sorts of toad. He had an option.

* Most species of toads mate in water, and during the mating period, the male clasps the female with his arms from a position on her back. Perhaps *"because"* she is slippery, he has roughened black pads on the dorsal sides of his hands in this season. In contrast, the midwife toad mates on land and has no such nuptial pads. In the years before World War I, Paul Kammerer, an Austrian scientist, claimed to have demonstrated the famous inheritance of acquired characters by forcing midwife toads to mate in water. Under these circumstances, the male developed nuptial pads. It was claimed that descendants of the male developed such pads, even on land.

LAMARCKIAN: But some of the descendants of the toad who made pads in water made pads even on land. That's what I mean by the inheritance of acquired characters.

BIOLOGIST: Ah, yes, I see. What was passed on was the loss of an option. The descendants could no longer breed normally on land. That's fascinating.

LAMARCKIAN: You are willfully failing to understand.

BIOLOGIST: Perhaps. But I still do not understand what is supposedly "passed on" or "inherited." The claimed empirical fact is that the descendants *differed* from the parent in lacking an option which the parent had. But this is not the passing on of a resemblance, which the word *inheritance* would suggest. It is the passing on of a *difference*. But the "difference" was not there to be passed on. The parent toad, as I understand it, still had his options in good shape.

And so on.

The crux of this argument is the logical typing of the genetic message that is supposed to be passed on. It is not good enough to say vaguely that the nuptial pads are passed on, and there is no point in claiming that the potentiality to develop nuptial pads is passed on because that potentiality was characteristic of the parent toad before the experiment began.*

Of course, it is not denied that the animals and to a lesser extent the plants in this world often present the appearance which we might expect in a world in which evolution had proceeded by pathways of Lamarckian inheritance.

We shall see that this appearance is inevitable given (a) that wild populations usually (perhaps always) are characterized by heterogeneous (mixed up and diverse) gene pools, (b) that individual animals are capable of somatic changes which are in some way adaptive, and (c) that mutation and the reshuffling of existing genes are random.

But this conclusion will follow only after the entropic economics of somatic change has been compared with the entropic economics of achieving the same phenotypic appearance by genetic determination.

* Arthur Koestler, in *The Case of the Midwife Toad* (New York: Vintage Books, 1973), records that at least one wild toad of this species has been found with nuptial pads. So the necessary genetic equipment is available. The evidential value of the experiment is seriously reduced by this finding.

In the imaginary dialogue, Lamarckian was silenced by the argument that the inheritance of acquired characteristics would be accompanied by loss of freedom to modify the individual body in response to the demands of habit or environment. This generalization is not quite so simply true. No doubt the substitution of genetic for somatic control (regardless of the question of heredity) will always diminish the flexibility of the individual. The option of somatic change in that particular characteristic will be wholly or partly lost. But the general question still remains: Does it *never* pay to substitute genetic for somatic control? If this were the case, the world would surely be a very different place from that which we experience. Likewise, if Lamarckian inheritance were the rule, the whole process of evolution and living would become tied up in the rigidities of genetic determination. The answer must be between these extremes, and lacking data that would untangle this matter, we are driven to common sense and deductions from cybernetic principles.

Let me illustrate the whole matter by a discussion of use and disuse.

2. USE AND DISUSE

This old pair of concepts, which used to be central in discussions of evolution, has almost dropped out of the argument, perhaps because in this connection it is especially necessary to keep clear the logical typing of the various components of any hypothesis.

That the effects of *use* might contribute in some way to evolution is not particularly mysterious. Nobody can deny that the biological scene looks, at a first glance, *as if* the effects of use and disuse were passed on from generation to generation. This, however, cannot be fitted into what we know of the self-corrective and adaptive nature of somatic change. The creatures would in very few generations lose all freedom of somatic adjustment.

But to go beyond the crude Lamarckian position is to face difficulties with the logical typing of the parts of the hypothesis. I believe these difficulties to be soluble. So far as *use* is concerned, it is not too difficult to think of sequences in which natural selection might favor those individuals whose genetic makeup would go along with the soma-

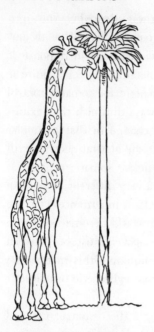

"Tell me, papa, why are the palm trees so tall?"
"It's so that the giraffes may be able to eat them, my child, for . . .

. . . if the palm trees were quite small, the giraffes would be in trouble (embarrassées)."

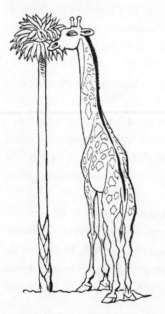

"But then, papa, why do the giraffes have such long necks?"
"Yes. It's so as to be able to eat the palm trees, my child, for . . .

. . . if the giraffes had short necks, they would be still more troubled."

tic changes current among the individuals of the given population. The somatic changes which accompany use are commonly (although not always) adaptive, and therefore genetic control which would favor such changes might be advantageous.

Under what circumstances does it pay, in terms of survival, to substitute genetic for somatic control?

The *price* of such a shift is, as I have argued, a lack of flexibility, but this lack must be spelled out more precisely if the conditions in which the shift will be beneficial are to be defined.

At first glance, there are those cases in which the flexibility would perhaps never be needed after the shift to the genetic. These are cases in which the somatic change is an adjustment to some *constant* environmental circumstance. Those members of a species that are settled in high mountains may as well base all their adjustments to mountain climate, atmospheric pressure, and the like on genetic determination. They do not need that reversibility which is the hallmark of somatic change.

Conversely, adaptation to variable and reversible circumstances is much better accomplished by somatic change, and it may well be that only very superficial somatic change is tolerable.

There is a grading of depth in somatic change. If a man goes up from sea level to 12,000 feet in the mountains, he will, unless he is in very good condition, begin to pant, and his heart will start to race. These immediate and reversible somatic changes are adequate to deal with a condition of emergency, but it would be an extravagant waste of flexibility to use panting and tachycardia as the ongoing adjustment to mountain atmosphere. What is required is somatic change which should be perhaps less reversible because we are now considering, not temporary emergency, but ongoing and lasting conditions. It will pay to sacrifice some reversibility in order to economize flexibility (i.e., to save the panting and tachycardia for some occasion in the high mountains when extra effort may be needed).

What will happen is called *acclimation*. The man's heart will undergo changes, his blood will come to contain more hemoglobin, his rib cage and respiratory habits will shift, and so on. These changes will be much less reversible than panting, and if the man goes visiting down in the plains, he will perhaps feel some discomfort.

In terms of the jargon of this book, there is a hierarchy of somatic adjustment dealing with particular and immediate demands at the superficial (most concrete) level and dealing with more general adjustment at deeper (more abstract) levels. The matter is exactly parallel to the hierarchy of learning in which protolearning deals with the narrow fact or action and deutero-learning deals with contexts and classes of context.

It is interesting to note that acclimation is accomplished by many changes on many fronts (heart muscle, hemoglobin, chest musculature, and so on); whereas the emergency measures tend to be ad hoc and specific.

What happens in acclimation is that the organism buys superficial flexibility at the price of deeper rigidity. The man can now use panting and tachycardia as emergency measures if he meets a bear, but he will be uncomfortable if he goes down to visit his old friends at sea level.

It is worthwhile to spell this matter out in more formal terms: Consider all the propositions that might be required to describe an organism. There may be millions of them, but they will be linked together in loops and circuits of interdependence. And in some degree, every descriptive proposition will be normative for that organism; that is, there will be maximum and minimum levels beyond which the variable described will be toxic. Too much sugar in the blood or too little will kill, and this is so for all biological variables. There is what can be called a *metavalue* attached to each variable; that is, it is good for the creature if the given variable is in the middle of its range, not at its maximum or minimum. And because the variables are interconnected in loops and circuits, it follows that any variable which is at maximum or minimum must partly cramp all other variables on the same loop.

Flexibility and survival will be favored by any change tending to keep variables floating in the middle of their range. But any extreme somatic adjustment will push one or more variables to extreme values. There is, therefore, always an available stress to be relieved by genetic change provided that the phenotypic expression of the change shall not be a further increase of already existing stress. What is required is a genetic change that will *alter the levels of tolerance for upper and/or lower values of the variable.*

If, for example, before genetic change (by mutation or, more probably, by reshuffling of genes), the tolerance for a given variable were within the limits 5 to 7, then a genetic change that would change the limits to a new value, 7 to 9, would have survival value for a creature whose somatic adjustment was straining to hold the variable up to the old value of 7. Beyond that, if the somatic adjustment pushed the new value to 9, there would be a further available increment of survival value to be gained by a further genetic change to permit or push the tolerance levels farther up the same scale.

In the past, it was difficult to account for evolutionary change related to *disuse*. That a genetic change in the same direction as the effects of habit or use would commonly have survival value was easy to imagine, but it was more difficult to see how a genetic duplication of effects of disuse might pay off. However, if the logical typing of the imagined genetic message is juggled, a hypothesis is achieved that uses a single paradigm to cover the effects of both use and disuse. The old mystery surrounding the blindness of cave animals and the eight-ounce femur of the eighty-ton blue whale is no longer totally mysterious. We have only to suppose that the maintenance of any residual organ, say a ten-pound femur in an eighty-ton whale, will always push one or more somatic variables to an upper or lower limit of tolerance to see that a shift of the limits of tolerance will be acceptable.

However, from the point of view of this book, this solution to the otherwise perplexing problems of use and disuse is an important illustration of the relation between genetic or somatic change and, beyond that, of the relation between higher and lower logical typing in the vast mental process called *evolution*.

The message of higher logical type (i.e., the more genetic injunction) does not have to mention the somatic variable whose tolerances are shifted by the genetic change. Indeed, the genetic script probably contains nothing in any way resembling the nouns or substantives of human language. My own expectation is that when the almost totally unknown realm of processes whereby DNA determines embryology is studied, it will be found that DNA mentions nothing but relations. If we should ask DNA how many fingers this human embryo will have, the answer might be, "Four paired relations between (fingers)." And if we ask how many gaps between fingers, the answer would be "three paired relations between (gaps)." In each case, only the *"relations between"* are defined and determined. The relata, the end components of the relationships in the corporeal world, are perhaps never mentioned.

(Mathematicians will note that the hypothetical system here described resembles their group theory in dealing only with relations among the *operations* by which something is transformed, never with the "something" itself.)

In this facet of the communication from somatic change through natural selection to the gene pool of the population, it is important to note

a. That somatic change is hierarchic in structure.

b. That genetic change is, in a sense, the highest component in that hierarchy (i.e., the most abstract and the least reversible).

c. That genetic change can at least partly avoid the price of imposing rigidity on the system by being delayed until it is probable that the circumstance which was coped with by the soma at a reversible level is indeed permanent and by acting only indirectly on the phenotypic variable. The genetic change presumably shifts only the *bias* or setting (see Glossary, "Logical Type") of the homeostatic control of the phenotypic variable.

d. That with this step from direct control of the phenotypic variable to control of the bias of the variable, there is also probably an opening and spreading of alternative possibilities for change. The control of tolerances for the size of the whale's femur is no doubt achieved by dozens of different genes acting, in this respect, together but each having perhaps quite other expressions in other parts of the body.

A similar breakdown from this single effect, in which the evolutionist happens to be interested at a given moment, to multiple alternative or synergistic causes was noted in the step from simple somatic change to acclimation. It is expectable that in biology, stepping from one logical level to the next higher will always have to be accompanied by this multiplication of relevant considerations.

3. GENETIC ASSIMILATION

What has been said in section 2 is exemplified in almost every point by my friend, Conrad Waddington's famous experiments demonstrating what he called *genetic assimilation*. The most dramatic of these began with the production of phenocopies of the effects on fruit flies of a gene called *bithorax*. All ordinary members of the vast order Diptera, except the wingless fleas, are two-winged and have the second pair of wings reduced to little rods with knobs at the ends that are believed to be balance organs. Under the gene bithorax, the wing rudiments in the third segment of the thorax become almost perfect wings, resulting in a four-winged fly.

This very profound modification of the phenotype, waking up very ancient and now inhibited morphology, could also be produced by a somatic change. When the pupae were intoxicated with ethyl ether in appropriate dosage, the adult flies, when they hatched, had the bithorax appearance. That is, the characteristic, bithorax, was known both as a product of genetics and as the product of violent disturbance of epigenesis.

Waddington performed his experiments on large populations of flies in big cages. In each generation, he subjected these populations to ether intoxication to produce the bithorax forms. In each generation, he selected out those flies that best represented his ideal of perfect bithorax development. (All were rather miserable-looking beasts, quite unable to fly.) From these selected individuals, he bred the next generation to be subjected to the ether treatment and selection.

From each generation of pupae, he kept a few before intoxication and let them hatch under normal conditions. Finally, as the experiment progressed, after some thirty generations, bithorax forms started to turn up in the untreated control group. Breeding from these showed that they were indeed produced, not by the single gene bithorax, but by a complex of genes that together create a four-winged appearance. In this experiment, there is no evidence of any direct inheritance of acquired characters. Waddington assumed that the shuffling of genes in sexual reproduction and the mutation rate were unaffected by the physiological insult to the organisms. What he offered as an explanation was that selection on an astronomical scale, perhaps eliminating from potential existence many tons of flies, sorted out a limited number of animals with bithorax. He argued that it was legitimate to see this as a selection of those individuals with the lowest threshold for the production of the bithorax anomaly.

We do not know what would have been the outcome of the experiment without Waddington's selecting of the "best" bithorax. Perhaps in thirty generations, he would have created a population immune to the ether treatment or conceivably a population needing ether. But perhaps if the bithorax modification was, like most somatic change, partly adaptive, the population would, like Waddington's experimental populations, have produced genetic copies (*genocopies*) of the results of ether treatment.

By the new word *genocopy,* I am stressing that the somatic change may, in fact, precede the genetic, so that it would be more appropriate to regard the genetic change as the copy. In other words, the somatic changes may partly determine the pathways of evolution; and this will be more so in larger gestalten than that which we are now considering. That is, we must again increase the logical typing of our hypothesis. Three steps in theory making can thus be distinguished:

a. At the individual level, environment and experience can induce somatic change but cannot affect the genes of the individual. There is no direct Lamarckian inheritance, and such inheritance *without selection* would irreversibly eat up somatic flexibility.

b. At the population level, with appropriate selection of phenotypes, environment and experience will generate better-adapted individuals on which selection can work. To this extent, the *population* behaves as a Lamarckian unit. It is no doubt for this reason that the biological world looks like a product of Lamarckian evolution.

c. But to argue that the somatic changes *pioneer* the direction of evolutionary change requires another level of logical typing, a still larger gestalt. We would have to invoke co-evolution and argue that the surrounding ecosystem or some closely abutting species will change to fit the somatic changes of the individuals. Such changes in environment could conceivably act as a mold which will favor any genocopy of the somatic changes.

4. THE GENETIC CONTROL OF SOMATIC CHANGE

Another aspect of the communication between genes and the development of the phenotype is disclosed when we ask about the genetic control of somatic change.

There is, surely, *always* a genetic contribution to all somatic events. I would argue as follows: If a man turns brown in the sun, we may say that this was a somatic change induced by exposure to light of the appropriate wavelengths and so on. If we subsequently protect him from the sun, the tanning he received will disappear, and if he is blond, he will get back his pinkish appearance. With further exposure to the sun, he will again go brown. And so on. The man changes color when exposed to sunshine, but his *ability* to change in this way is not affected by the exposure to or the protection from the sun—or so I believe.

But it is conceivable (and in the more complex processes of learning, it is evidently so) that the *ability* to achieve certain somatic changes is subject to learning. It is as if the man could improve or reduce his ability to tan under sunlight. In such a case, the ability to achieve this metachange might be totally controlled by genetic factors. Or it is conceivable that, again, there might be an ability to *change the ability to change.* And so on. But in no real case is it possible that the series of steps could be infinite.

It follows that the series must always end up in the genome, and it seems probable that in most instances of learning and somatic change, the number of levels of somatic control is small. We can learn and learn to learn and possibly learn to learn to learn. But that is probably the end of the sequence.

On the basis of these considerations, it is nonsensical to ask: Is the given characteristic of that organism determined by its genes or by somatic change or learning? There is no phenotypic characteristic that is unaffected by the genes.

The more appropriate question would be: At what level of logical typing does genetic command act in the determining of this characteristic? The answer to this question will always take the form: At one logical level *higher* than the observed ability of the organism to achieve learning or bodily change by somatic process.

Because of this failure to recognize logical typing of genetic and of somatic change, almost all comparisons of "genius," inherited "capacity," and the like degenerate into nonsense.

5. "NOTHING WILL COME OF NOTHING" IN EPIGENESIS

I have already pointed out that epigenesis is to evolution as the working out of a tautology is to creative thought. In the embryology of a creature, not only is there no need for new information or change of plan, but for the most part, epigenesis must be protected from the intrusion of new information. The way to do it is the way it has always been done. The development of the fetus should follow the axioms and postulates laid down in DNA or elsewhere. In the language of Chapter 2, evolution and learning are necessarily *divergent* and unpredictable, but epigenesis should be convergent.

It follows that in the field of epigenesis, the cases in which new information is needed will be rare and conspicuous. Conversely, there should be cases, albeit pathological, in which lack or loss of information results in gross distortions of development. In this context, the phenomena of symmetry and asymmetry become a rich hunting ground in which to look for examples. The ideas that must guide the early embryo in these respects are both simple and formal, so that their presence or absence is unmistakable.

The best-known examples come from the experimental study of the embryology of amphibians, and I shall discuss here some of the phenomena connected with symmetry in the frog's egg. What is known of the frog is probably true of all vertebrates.

It seems that without information from the outside world, the unfertilized frog's egg does not contain the necessary information (i.e., the necessary *difference*) to achieve bilateral symmetry. The egg has two differentiated poles: the *animal* pole, where protoplasm preponderates over yolk, and the *vegetal* pole, where yolk is preponderant. But there is no differentiation among meridians or lines of longitude. The egg is in this sense radially symmetrical.

No doubt the differentiation of animal and vegetal poles was determined by the position of the egg in the follicular tissue or by the plane of the last cell division in gamete production; that plane, in turn, was probably determined by position of the mother cell in the follicle. But this is not enough.

Without some differentiation among the sides or meridians of the unfertilized egg, it is impossible for the egg to "know" or "decide" which shall be the future median plane of symmetry of the bilaterally symmetrical frog. Epigenesis cannot begin until one meridian is made different from all others. Fortunately, we happen to know how this crucial information is provided. It comes, necessarily, from the outside world and is the entry point of the spermatozoon. Typically, the spermatozoon enters the egg somewhat below the equator, and the meridian that includes the two poles and the entry point defines the median plane of the frog's bilateral symmetry. The first segmentation of the egg follows that meridian, and the side of the egg on which the spermatozoon enters becomes the ventral side of the frog.

Furthermore, it is known that the needed message is not carried in DNA or other complexities of the structure of the spermatozoon. A prick with the fiber of a camel's hair brush will do the trick. Following such a prick, the egg will segment and continue development, becoming an adult frog that will hop and catch flies. It will, of course, be haploid (i.e., will lack half the normal complement of chromosomes). It will not breed, but it will otherwise be perfect in all respects.

A spermatozoon is not necessary for this purpose. All that is needed is a *marker of difference,* and the organism is not particular regarding the character of this marker. Without some marker, there will be no embryo. "Nothing will come of nothing."

But this is not the end of the story. The future frog and, indeed, already the very young tadpole is conspicuously asymmetrical in its endodermic anatomy. Like most vertebrates, the frog is rather precisely symmetrical in ectoderm (skin, brain, and eyes) and in mesoderm (skeleton and skeletal muscles) but is grossly asymmetrical in its endodermic structures (gut, liver, pancreas, and so on). (Indeed, every creature that folds its gut in other than the median plane must be asymmetrical in this respect. If you look at the belly of any tadpole, you will see the gut, clearly visible through the skin, coiled in a great spiral.)

Expectably, situs inversus (the condition of reversed symmetry) occurs among frogs, but with extreme rarity. It is well known in the human species and affects about one individual in a million. Such individuals look just like other people but internally they are reversed, the right side of the heart serving the aorta while the left serves the lungs, and so on. The causes of this reversal are not known, but the fact that it occurs at all denotes that the normal asymmetry is *not* determined by the asymmetry of the molecules. To reverse any part of that chemical asymmetry would require the reversal of all because the molecules must appropriately fit each other. Reversal of the entire chemistry is unthinkable and could not survive in any but a reversed world.

So a problem remains regarding the source of the information which determines the asymmetry. There must surely be information that will instruct the egg with regard to the correct (statistically normal) asymmetry.

So far as we know, there is no moment after fertilization at which this information could be delivered. The order of events is first extrusion from the mother, then fertilization; after that, the egg is protected in a mass of jelly throughout the period of segmentation and early embryonic development. In other words, the egg must surely already contain the information necessary to determine asymmetry *before* fertilization. In what form must this information exist?

In the discussion of the nature of explanation in Chapter 2, I noted that no dictionary can define the words *left* or *right*. That is, no arbitrary digital system can resolve the matter; the information must be ostensive. We now have the chance of finding out how the same problem is solved by the egg.

I believe that there can be, in principle, only one sort of solution (and I hope that somebody with a scanning electron microscope will look for the evidence). It must be so that the answer is in the egg before fertilization and therefore is in some form that will still determine the same asymmetry *regardless of which meridian is marked by the entering spermatozoon*. It follows that every meridian, regardless of where it is drawn, must be asymmetrical and that all must be asymmetrical in the same sense.

This requirement is satisfied most simply by some sort of *spiral of nonquantitative or vector relations*. Such a spiral will cut every meridian obliquely to make in every meridian the same difference between east and west.

A similar problem arises in the differentiation of bilateral limbs.

My right arm is an asymmetrical object and a formal mirror image of my left. But there are in the world rare monstrous individuals who bear a pair of arms or a forked arm on one side of the body. In such cases, the pair will be a bilaterally symmetrical system. One component will be a right and one a left, and the two will be so placed as to be in mirror image.* This generalization was first enunciated by my father in the 1890s and for a long time was called *Bateson's rule.* He was able to show the working of this rule in almost every phylum of animals by a search of all the museums and many primate collections in Europe and America. Especially, he gathered about a hundred cases of such aberration in the legs of beetles.

I reexamined this story and argued, from his original data, that he had been wrong to ask: What determined this extra symmetry? He should have asked: What determined the *loss* of asymmetry?

I proposed the hypothesis that the monstrous forms were produced by *loss or forgetting* of information. To be bilaterally symmetrical requires more information than radial symmetry, and to be asymmetrical requires more information than bilateral symmetry. Asymmetry of a lateral limb, such as a hand, requires appropriate orientation in three directions. The direction towards the back of the hand must be different from the direction towards the palm; the direction towards the thumb must be different from the direction towards the little finger, and the direction towards the elbow must be different from the direction towards the fingers. These three directions must be appropriately put together to make a *right* rather than a *left* hand. If one direction is reversed, as when the hand is reflected in a mirror, a reversed image will result (see Chapter 3, section 9). But if one of the three differentiations is *lost or forgotten,* the limb will be able to achieve only bilateral symmetry.

In this case, the postulate "nothing will come of nothing" becomes a little more elaborate: Bilateral symmetry will come of asymmetry when one discrimination is lost.

---

*I have simplified the rule somewhat for this presentation. For a more complete account see *Steps to an Ecology of Mind* in the essay entitled "A Re-examination of Bateson's Rule."

6. HOMOLOGY

At this point, I shall leave the problems of individual genetics, somatic change and learning, and the immediate pathways of evolution to look at the results of evolution on the larger scale. I shall ask what we can deduce about the underlying processes from the wider picture of phylogeny.

Comparative anatomy has a long history. For at least sixty years, from the publication of *The Origin of Species* to the 1920s, the focus of comparative anatomy was on relatedness, to the exclusion of process. That phylogenic trees could be constructed was felt to be evidence for Darwinian theory. The fossil record was inevitably very incomplete, and lacking such direct evidence of descent, the anatomists showed an insatiable appetite for instances of that class of resemblances called *homology*. Homology "proved" relatedness, and relatedness was evolution.

Of course, people had noted the formal resemblances among living things at least since the evolution of language, which classified my "hand" with your "hand" and my "head" with the "head" of a fish. But awareness of any need to explain such formal resemblances came much later. Even today, most people are not surprised by, and see no problem in, the resemblance between their two hands. They do not feel or see any need for a theory of evolution. To the thoughtful among the ancients and even to people of the Renaissance, the formal resemblance between creatures illustrated the connectedness within the Great Chain of Being, and these connections were logical, not genealogical, links.

Be all that as it may, the jumped conclusion from formal resemblance to relatedness concealed a number of jumped hypotheses.

Let us grant the formal resemblance in thousands of cases—man and horse, lobster and crab—and let us assume that in these cases, the formal resemblances are not merely evidence for but flatly *the result of* evolutionary relationship. We can then go on to consider whether the nature of the resemblances in these cases throws light on the evolutionary process.

We ask: What do the homologies tell us about the *process* of evolution? What we find, when we compare our description of lobster with our description of crab, is that some components of the descriptions remain unchanged and that others are different from one description to the other. Therefore, our first step will surely take the form of distinguishing between different sorts of change. Some changes will be stressed as more probable and easy; others will be more difficult and therefore more improbable. In such a world, the slow-changing variables will lag behind and could become the core of those homologies on which the wider hypotheses of taxonomy might be based.

But this first classification of changes into *fast* and *slow* will itself require explanation. What can we add to our description of evolutionary process that will, perhaps, let us predict which variables will, in fact, be slow changing and so become the basis of homology?

To my knowledge, the only beginning of such a classification is implicit in the theory of so-called recapitulation.

The germ of the doctrine of recapitulation was first put forward by the early German embryologist, Karl Ernst von Baer in 1828 in the phrase "law of corresponding stages." He demonstrated his law by the device of comparing unlabeled vertebrate embryos.

> I am quite unable to say to what class they belong. They may be lizards or small birds or very young mammalia, so complete is the similarity in the mode of formation of the head and trunk in these animals. The extremities are still absent, but even if they existed, in the earlier stage of development we should learn nothing because all arise from the same fundamental form.*

Von Baer's concept of "corresponding stages" was later elaborated by Ernst Haeckel, Darwin's contemporary, into the theory of recapitulation and the much-disputed assertion that "ontogeny repeats phylogeny." Since then, very varied phrasings of the matter have been proposed. Most cautious is perhaps the assertion that the larvae or embryos of a given species commonly resemble the *larvae* of a related species more closely than the adults resemble the adults of the related species. But even this very cautious phrasing is blemished by conspicuous exceptions.†

However, in spite of the exceptions, I incline to the view that

---

Encyclopedia Britannica, S.V. "Baer, Karl Ernst von (1792–1876)."

†For example, among the marine wormlike creatures of the older Enteropneusta, different species, of what used to be regarded as a single genus *Balanoglossus,* have totally different embryology. *B. kovalevskii,* has tadpolelike larvae with gill slits and notochord; whereas other species have larvae like those of echinoderms.

von Baer's generalization provides an important clue to evolutionary process. Right or wrong, his generalization poses important questions about the survival not of organisms but of traits: Is there any highest common factor shared by those variables that become stable and therefore have been used by zoologists in their search for homology? The law of corresponding stages has an advantage over later phrasings in that he was not grasping after phylogenic trees, and even the brief quotation I have cited has special points that would be unnoticed by a phylogenetic sleuth. Is it so that embryonic variables are more enduring than adult variables?

Von Baer is concerned with higher vertebrates: lizards, birds, and mammals, creatures whose embryology is padded and protected either by an eggshell full of food or in a womb. With, say, insect larvae, von Baer's demonstration simply would not work. Any entomologist could look at an unlabeled display of beetle larvae and say at once to what family each larva belongs. The diversity of the larvae is as surprising as the diversity of the adults.

The law of corresponding stages is seemingly true not only of whole vertebrate embryos but also of successive limbs in the earliest stages of their development. So-called serial homology shares with phylogenic homology the generalization that, on the whole, *resemblances precede differences.* The fully developed claw of a lobster differs conspicuously from the walking appendages on the other four thoracic segments, but all the thoracic appendages looked alike in their early stages.

Perhaps this is as far as we should push von Baer's generalization: to assert that, in general, resemblance is *older* (both phylogenetically and ontogenetically) than difference. To some biologists, this will seem like a truism, as if to say that in any branching system, two points close to the point of branching will be more like each other than will two points far from that point. But this apparent truism would not be true of elements in the periodic table and would not necessarily be true in a biological world produced by special creation. Our truism is, in fact, evidence for the hypothesis that organisms are indeed to be related as points or positions on a branching tree.

The generalization that resemblance is older than difference is still a very incomplete explanation of the occurrence of homology in thousands of examples throughout the biological world. The question, "why do some characteristics become the basis of homology?" is only repeated by saying that resemblances are older than differences. The question remains almost unchanged: Why do some characteristics become older, surviving longer, to become the basis of homology?

We face a problem of *survival*, not the survival of species or varieties struggling in a hostile world of other organisms, but a more subtle survival of *traits* (items of description) that must survive both in an outside environment and in an inner world of other traits in the total reproduction, embryology, and anatomy of the given organism.

In the complex network of the scientist's description of the total organism, why do some pieces of that description stay true longer (over more generations) than other pieces? And is there coincidence, overlapping, or synonymy between the parts and pieces of the description and the parts and pieces of the aggregate of injunctions that determine ontogenesis?

If an elephant had the dentition and other formal characteristics of members of the family Muridae, he would be a mouse in spite of his size. Indeed, the cat-sized hyrax is close to being a hippopotamus, and the lion is very close to being a pussycat. Mere size seems to be irrelevant. Form is what matters. But what precisely is meant by "form" or "pattern" in this context is not easy to define.

We are searching for criteria whereby we can recognize those traits that are appropriate candidates for ongoing truth in the hurly-burly of evolutionary process. Two characteristics of such traits stand out—two traditional ways of dividing up the vast field of "differences." There is the dichotomy between pattern and quantity and the dichotomy between continuity and discontinuity. Are contrasting organisms linked by a continuous series of steps, or is there a sudden transition between them? It is awkward (but not impossible) to imagine gradual transition between patterns, and therefore, these two dichotomies are likely to overlap. At the very least, we can expect that those theorists who prefer to invoke pattern will also prefer theories that invoke discontinuity. (But, of course, such preferences, which depend only upon the bent of the mind of the individual scientist or the fashionable opinions of the time, should be deprecated.)

The clearest findings relevant to this subject are, I believe, the elegant demonstrations of the zoologist D'Arcy Wentworth Thompson in the early part of this century. He showed that in many cases, perhaps in every case he tested, two contrasting but related animal forms will have this in common: that if one form is drawn (say, in profile) on simple orthogonal Cartesian coordinates (e.g., on squared paper), then with appropriate bending or distortion, the same coordinates will accommodate the other form. All points on the profile of the second form will fall on similarly named points on the bent coordinates. (See Figure 9.)

What is significant in D'Arcy Thompson's findings is that in any given case, the distortion is unexpectedly simple and is consistent throughout the depiction of the animal. The bending of the coordinates is such as could be described by some simple mathematical transformation.

This simplicity and consistency must surely mean that those *differences* between the phenotypes, which D'Arcy Thompson's method exposes, are represented by rather few differences of genotype (i.e., by rather few genes).

Furthermore, from the consistency of distortion throughout the animal's body, it would seem that the genes in question are pleiotropic (i.e., affect many, perhaps *all,* parts of the phenotype) in ways that are, in this particular sense, harmonious throughout the body.

To interpret these findings any further is not entirely simple, and D'Arcy Thompson himself does not do much to help. He is exultant that mathematics is shown able to describe certain sorts of change.

In this connection, it is interesting to note the current controversy between the upholders of "synthetic" theory in evolution (the current Darwinian orthodoxy) and their enemies, the "typologists." Ernst Mayr, for example, makes fun of the blindness of typologists: "History shows that the typologist does not and cannot have any appreciation of natural selection."* Unfortunately, he does not quote his sources for his typology of his colleagues. Is he too modest to claim the credit? Or is it so, in this case, that it takes one to know one?

Are we not all typologists under the skin?

In any case, there are no doubt many ways of looking at animal forms. And because we are embarked on a Platonic study of the parallelism between creative thinking and that vast mental process called *biologi-*

* See Ernst Mayr, *Populations, Species and Evolution* (Cambridge, Mass.: Harvard University Press, 1963), p. 107.

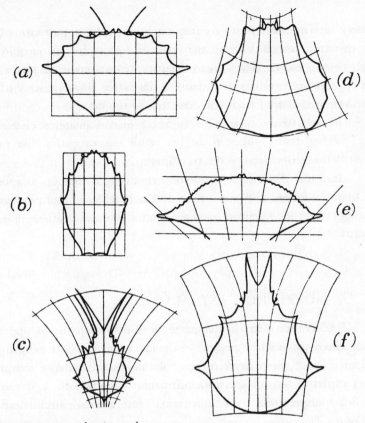

Figure 9. Carapaces of various crabs.

This figure reproduced from D'Arcy Thompson's *On Growth and Form*, p. 294. Reprinted by permission of Cambridge University Press, copyright © 1961.

cal evolution, it is worthwhile to ask in every instance: Is *this* way of looking at the phenomena somehow represented or paralleled within the organizational system of the phenomena themselves? Do any of the genetic messages and static signs that determine the phenotype have the sort of syntax (for lack of a better word) which would divide "typological" from "synthetic" thinking? Can we recognize, among the very messages which create and shape the animal forms, some messages more typological and some more synthetic?

When the question is put in this way, it seems that Mayr is deeply *right* in proposing his typology. The old drawings of D'Arcy Thompson precisely *separate* two sorts of communication within the organism itself. The drawings show that animals have two sorts of characteristics: (a) They have relatively stable quasi-topological patterns, which have understandably led scientists to postulate gross discontinuity in the evolutionary process. These characteristics remain constant under the impact of (b) the relatively unstable quantitative characteristics which are shown as changing from one depiction to the next.

If we draw the coordinates to fit the quasi-topological character-istics, we find that changes in the less stable characteristics have to be represented as distortions of the coordinates.

In terms of the present question regarding homology, it appears that there are indeed different sorts of characteristics and that phylogenic homology will surely depend upon the more stable and quasi-topological patterns.

7. ADAPTATION AND ADDICTION

"Adaptation" in the language of the evolutionist is approxi-mately synonymous with "design" in the language of such theologians as William Paley,* whose *Evidences* is a voluminous collection of conspicu-ous examples of elegant special adjustments of animals to their way of life. But I suspect that both "adaptation" and "design" are misleading concepts.

If we come to regard the production of particular pieces of adap-tation—the claw of the crab, the hand and eye of the man, and so on—as central to the mass of problems the evolutionist must solve, we distort and limit our view of evolution as a whole. What seems to have happened, perhaps as a result of the silly battles between the early evolu-tionists and the Church, is that out of the vast Heraclitean flux of evolu-tionary process, certain eddies and backwaters of the stream have been picked out for special attention. As a result, the two great stochastic processes have been partly ignored. Even professional biologists have not seen that in the larger view, evolution is as value-free and as beautiful as the dance of Shiva, where all of beauty and ugliness, creation and de-struction are expressed or compressed into one complex symmetrical pathway.

* William Paley (1743–1805) was a defender of the Genesis story of creation long before Darwin was born. His *View of the Evidences of Christianity* (1794) was until recently a required subject for those Cambridge students who did not take Greek.

By setting the terms *adaptation* and *addiction* side by side in the title of this section, I have tried to correct this sentimental or at least overoptimistic view of evolution as a whole. The fascinating cases of adaptation which make nature appear so clever, so ingenious, may also be early steps toward pathology and overspecialization. And yet it is difficult to see the crab's claw or the human retina as first steps toward pathology.

It seems that we must ask: What characterizes those adaptations that turn out to be disastrous, and how do these differ from those that seem to be benign and, like the crab's claw, remain benign through geological ages?

The question is pressing and relevant to the contemporary dilemmas of our own civilization. In Darwin's day, every invention appeared benign, but that is not so today. Sophisticated eyes in the twentieth century will view every invention askance and will doubt that blind stochastic processes always work together for good.

We badly need a science that will analyze this whole matter of adaptation-addiction at all levels. Ecology is perhaps the beginning of such a science, although ecologists are still far from telling us how to get out of an atomic armaments race.

In principle, neither random genetic change accompanied by natural selection nor random processes of trial and error in thought accompanied by selective reinforcement will necessarily work for the good of either species or individual. And at the social level, it is still not clear that the inventions and stratagems which are rewarded in the individual necessarily have survival value for the society; nor, vice versa, do the policies that representatives of society might prefer necessarily have survival value for individuals.

A large number of patterns can be adduced which suggest that a belief in natural selection or *laissez-faire* is obviously naïve:

a. The remainder of the system will change to crowd in on the innovation to make it irreversible.

b. Interaction with other species or individuals will lead to a change in context, so that further innovation of the same kind becomes necessary, and the system then goes into escalation or runaway.

c. The innovation sets up other changes within the system, making it necessary to forgo other adaptations.

d. The flexibility (i.e., positive entropy) of the system is eaten up.

e. The adapted species is so favored that by overgrazing in some form, it will destroy its ecological niche.

f. What seemed desirable in short time perspective becomes disastrous over longer time.

g. The innovating species or individual comes to act as if it is no longer partially dependent on neighboring species and individuals.

h. By a process of addiction, the innovater becomes hooked into the business of trying to hold constant some rate of change. The social addiction to armaments races is not fundamentally different from individual addiction to drugs. Common sense urges the addict always to get another fix. And so on.

In sum, each of these disasters will be found to contain an error in logical typing. In spite of immediate gain at one logical level, the sign is reversed and benefit becomes calamity in some other, larger or longer, context.

We lack any systematic knowledge of the dynamics of these processes.

8. STOCHASTIC, DIVERGENT, AND CONVERGENT PROCESSES

Ross Ashby* long ago pointed out that no system (neither computer nor organism) can produce anything *new* unless the system contains some source of the random. In the computer, this will be a random-number generator which will ensure that the "seeking," trial-and-error moves of the machine will ultimately cover all the possibilities of the set to be explored.

In other words, all innovative, *creative* systems are, in the language of Chapter 2, *divergent;* conversely, sequences of events that are predictable are, ipso facto, convergent.

* See W. Ross Ashby, *Introduction to Cybernetics.* (New York and London: John Wiley and Sons, Inc., 1956).

This does not mean, by the way, that all divergent processes are stochastic. For that, the process requires not only access to the random but also a built-in comparator that in evolution is called "natural selection" and in thought "preference" or "reinforcement."

It may well be that under the eye of eternity, which sees everything in cosmic and eternal context, *all* event sequences become stochastic. To such an eye, and even to the patient and compassionate Taoist saint, it may be clear that no ultimate preference is necessary for the steering of the total system. But we live in a limited region of the universe, and each one of us exists in limited time. To us, the divergent is real and is a potential source of either disorder or innovation.

I even suspect sometimes that we, albeit bound in illusion, do the Taoist's work of choosing and preferring while he sits back. (I am reminded of the mythical poet who was also a conscientious objector. He claimed, "I am the civilization for which the other boys are fighting." Perhaps he was, in some sense, right?)

Be all that as it may, it appears that we exist in a limited biosphere whose major bent is determined by two interlocking stochastic processes. Such a system cannot long remain without change. But the *rate* of change is limited by three factors:

a. The Weissmannian barrier between somatic and genetic change, discussed in section 1 of this chapter, ensures that the somatic adjustments shall not rashly become irreversible.

b. In every generation, sexual reproduction provides a guarantee that the DNA blueprint of the new shall not conflict outrageously with the blueprint of the old, a form of natural selection operating at the level of DNA regardless of what the deviant new blueprint may mean to the phenotype.

c. Epigenesis operates as a convergent and conservative system; the developing embryo is, within itself, a context of selection favoring conservatism.

It was Alfred Russel Wallace who saw clearly that natural selection is a conservative process. His quasi-cybernetic model, in his letter explaining his idea to Darwin, has been mentioned elsewhere but is relevant here:

The action of this principle is exactly like that of the centrifugal governor of the steam engine, which checks and corrects any irregularities almost before they become evident; and in like manner no unbalanced deficiency in the animal kingdom can ever reach any conspicuous magnitude, because it would make itself felt at the very first step, by rendering existence difficult and extinction almost sure to follow.*

9. COMPARING AND COMBINING THE TWO STOCHASTIC SYSTEMS†

In this section, I shall try to make more precise the description of the two systems, to examine the functions of each, and finally, to examine the character of the larger system of total evolution that is the product of combining the two subsystems.

Each subsystem has two components (as is implied by the word *stochastic*) (see Glossary): a random component and a process of selection working on the products of the random component.

In that stochastic system to which Darwinians have paid most attention, the random component is *genetic* change, either by mutation or by the reshuffling of genes among members of a population. I assume mutation to be nonresponsive to environmental demand or to internal stress of the organism. I assume, however, that the machinery of selection which acts on the randomly varying organisms will include both each creature's internal stress and, later, the environmental circumstances to which the creature is subjected.

It is of primary importance to note that insofar as embryos are protected in eggs or in the mother's body, the external environment will not have a strong selective effect on genetic novelties until epigenesis has proceeded through many steps. In the past and still continuing into the present, external natural selection has favored those changes that protect the embryo and juvenile from external dangers. The result has been an increasing separation between the two stochastic systems.

* See Alfred Russel Wallace, "On the Tendency of Varieties to Depart Indefinitely from the Original Type," Linnaean Society Papers (London, 1858). Reprinted in P. Appleman, ed., *Darwin, A Norton Critical Edition* (New York: W. W. Norton, 1970), p. 97.
† This section is the most difficult and perhaps the most important part of the book. The lay reader and especially the reader who needs to see the *usefulness* of all thinking will perhaps find help in Appendix I, which reproduces a memorandum addressed to the regents of the University of California.

An alternative method for ensuring the survival of at least a few of the offspring is by vast multiplication of their numbers. If every reproductive cycle of the individual produces *millions* of larvae, the rising generation can suffer decimation some six times over. This is to treat the external causes of death as probabilistic, making no attempt to adapt to their particular nature. By this strategy, too, the internal selection is given a clear field for the control of change.

Thus, either by protection of the immature offspring or by their astronomical multiplication, it comes about that today, for many organisms, the internal conditions will provide the *first* constraint to which the new form must conform. Will the new form be viable in that setting? Will the developing embryo be able to tolerate the new form, or will the change precipitate lethal irregularities in the embryo's development? The answer will depend upon the somatic flexibility of the embryo.

Above all, in sexual reproduction, the matching up of chromosomes in fertilization enforces a process of comparison. What is new in either ovum or spermatozoon must meet with what is old in the other, and the test will favor conformity and conservation. The more grossly new will be eliminated on grounds of incompatibility.

Following the fusion process of reproduction will come all the complexities of development, and here the combinatorial aspect of embryology which is stressed in the term *epigenesis* will impose further tests of conformity. We know that in the status quo ante, all the requirements of compatibility were met to produce a sexually mature phenotype. If this were not so, the status quo ante could never have existed.

It is very easy to fall into the notion that if the new is viable, then there must have been something wrong with the old. This view, to which organisms already suffering the pathologies of over rapid, frantic social change are inevitably prone, is, of course, mostly nonsense. What is *always* important is to be sure that the new is not *worse* than the old. It is still not certain that a society containing the internal combustion engine can be viable or that electronic communication devices such as television are compatible with the aggressive intraspecies competition generated by the Industrial Revolution. Other things being equal (which is not often the case), the old, which has been somewhat tested, is more likely to be viable than the new, which has not been tested at all.

Internal selection, then, is the first maze of tests to which any new genetic component or combination is subject.

In contrast, the second stochastic system has its immediate roots in external adaptation (i.e., in the interaction between phenotype and environment). The random component is provided by the system of phenotype in interaction with environment.

The particular acquired characteristics produced in response to some given change in environment may be predictable. If the food supply is reduced, the organism is likely to lose weight mainly by metabolizing its own fat. Use and disuse will bring about changes in the development or underdevelopment of particular organs. And so on. Similarly, within the environment, prediction of particular change is often possible: a change of climate toward greater cold may predictably reduce the local biomass and so reduce the food supply for many species of organisms. But *together,* the phenotype and the organism generate an unpredictability.* Neither organism nor environment contains information about what the other will do next. But in this subsystem, a selective component is already present insofar as somatic changes evoked by habit and environment (including habit itself) are adaptive. (*Addiction* is the name of the large class of changes induced by environment and experience that are not adaptive and do not confer survival value.)

Between them, environment and physiology *propose* somatic change that may or may not be viable, and it is the current state of the organism as determined by *genetics* that determines the viability. As I argued in section 4, the limits of what can be achieved by somatic change or by learning are always ultimately fixed by genetics.

In sum, the combination of phenotype and environment thus constitutes the random component of the stochastic system that *proposes* change; the genetic state *disposes,* permitting some changes and prohibiting others. Lamarckians want the somatic change to control the genetic, but in truth, the opposite is the case. It is the genetics that limits the somatic changes, making some possible and some impossible.

Moreover, as that which contains potentials for change, the genome of the individual organism is what the computer engineers would

---

* The reader may be interested in comparing this unpredictability, generated by these two interacting subsystems, with the unpredictability generated by the interaction of Alice and her flamingo in the famous game of croquet.

call a *bank,* providing storage of available alternative pathways of adaptation. Most of these alternatives remain unused and therefore invisible in any given individual.

Similarly, in the other stochastic system, the gene pool of the *population* is nowadays believed to be exceedingly heterogeneous. All of the genetic combinations that could occur are created, if only rarely, by the shuffling of genes in sexual reproduction. There is thus a vast bank of alternative genetic pathways that any wild population can take under pressure of selection, as is shown in Waddington's studies of genetic assimilation (discussed in section 3).

So far as this picture is correct, both population and individual are ready to move. There is, expectably, no need to wait for appropriate mutations, which is a point of some historic interest. Darwin, as is well known, shifted his views about Lamarckism in the belief that geological time was insufficient for a process of evolution which would operate without Lamarckian inheritance. He therefore accepted a Lamarckian position in later editions of *The Origin of Species.* Theodosius Dobzhansky's discovery that the unit of evolution is the population and that the population is a heterogeneous storehouse of genic possibilities greatly reduces the time required by evolutionary theory. The population is able to respond immediately to environmental pressures. The individual organism has the capacity for adaptive somatic change, but it is the population that, by selective mortality, undergoes change which is transmitted to future generations. The *potentiality* for somatic change becomes the object of selection. It is on *populations* that environmental selection acts.

We now proceed to examine the separate contributions of each of these two stochastic systems to the overall process of evolution. Clearly, in each case, it is the selective component that gives direction to the changes which are finally incorporated into the total picture.

The time structure of the two stochastic processes is necessarily different. In the case of random genetic change, the new state of DNA is in existence from the moment of fertilization but will perhaps not contribute to external adaptation until much later. In other words, the first test of genetic change is *conservative.* It follows that it is this internal stochastic system which will ensure that formal resemblance in internal relations between parts (i.e., homology) will be conspicuous everywhere. In addition, it is possible to predict which among the many sorts of homology will be most favored by internal selection; and the answer is *first* the cytological, that most surprising set of resemblances which unites the whole world of cellular organisms. Wherever we look, we find comparable forms and processes within the cells. The dance of the chromosomes, the mitochondria and other cytoplasmic organelles, and the uniform ultramicroscopic structure of flagella wherever they occur, either in plants or in animals—all these very profound formal resemblances are the result of internal selection that insists on conservatism at this elementary level.

A similar conclusion emerges when we ask about the later fate of changes that have survived the first cytological tests. The change that has impact *earlier* in the life of the embryo must disturb a longer and correspondingly more complex chain of later events.

It is difficult or impossible to establish any quantitative estimate of the distribution of homologies through the life history of the creatures. To assert that homology is most prevalent at very early stages in gamete production, fertilization, and so on is to make a quantitative statement identifying *degrees* of homology, setting a value on such characteristics as chromosome number, mitotic pattern, bilateral symmetry, five-toed limbs, dorsal central nervous systems, and so on. Such evaluation will be very artificial in a world in which (as noted in Chapter 2) quantity never determines pattern. But the hunch still remains. The *only* formal patterns shared by all cellular organisms—plants and animals alike—are at the cellular level.

An interesting conclusion follows from these lines of thought: After all the controversy and skepticism, the theory of recapitulation is defensible. There is a priori reason to expect that embryos will resemble in formal pattern the embryos of ancestral forms more closely than the formal patterns of adults will resemble those of ancestral adults. This is far from what Haeckel and Herbert Spencer dreamed of in their notion that embryology would have to follow the pathways of phylogeny. The present phrasing is more negative: Deviation from the beginning of the pathway is more difficult (less probable) than deviation from later stages.

If, as evolutionary engineers, we faced the task of choosing a pathway of phylogeny from free-swimming, tadpolelike creatures to the sessile, wormlike *Balanoglossus* living in mud, we would find that the easiest course of evolution would avoid too early and too drastic disturbances of the embryologic stages. We might even find that it would be a simplification of *evolutionary* process to punctuate epigenesis by a demarcation of separate stages. We would then arrive at a creature with free-swimming, tadpolelike larvae that, at a certain moment, would undergo metamorphosis into the wormlike, sessile adults.

The machinery of change is not simply permissive or simply creative. Rather, there is a continual determinism whereby the changes that can occur are members of a *class* of changes appropriate to that particular machinery. The system of random genetic change filtered by the selective process of internal viability gives to phylogeny the characteristic of pervasive homology.

If we now consider the other stochastic system, we shall arrive at a quite different picture. Although no learning or somatic change can directly affect DNA, it is clearly so that somatic changes (i.e., the famous acquired characteristics) are commonly adaptive. It is useful in terms of individual survival and/or reproduction and/or simple comfort and stress reduction to adjust to environmental change. Such adjustment occurs at many levels, but at every level, there is a real or seeming benefit. It is a good idea to pant when you arrive at a high altitude and a good idea to learn not to pant if you stay long in the high mountains. It is a good idea to have a physiological system that will adjust to physiological stress, even though adjustment leads to acclimation and acclimation may be addiction.

In other words, somatic adjustment will always create a context for genetic change, but whether such genetic change will follow is a quite separate question. Let me set that question aside for the moment and consider the spectrum of what *can* be proposed by somatic change. Clearly, this spectrum or set of possibilities will set an outward limit to what this stochastic component of evolution can achieve.

One common characteristic of somatic change is immediately evident: *All* such changes are *quantitative* or, as the computer engineers would say, *analogic*. In the animal body, the central nervous system and

DNA are in large degree (perhaps totally) digital, but the remainder of the physiology is analogic.*

Thus, in comparing the random genetic changes of the first stochastic system with the responsive somatic changes of the second, we meet again with the generalization stressed in Chapter 2: *Quantity does not determine pattern.* The genetic changes may be highly abstract, operating at many removes from their ultimate phenotypic expression, and no doubt, they may be either quantitative or qualitative in their final expression. But the somatic are much more direct and are, I believe, solely quantitative. The descriptive propositions that contribute shared pattern (i.e., homology) to the description of species are, so far as I know, never disturbed by the somatic changes that habit and environment can induce.

In other words, the contrast that D'Arcy Thompson demonstrated (see Figure 9) would seem to have roots in (i.e., to follow from) this contrast between the two great stochastic systems.

Finally, I have to compare the processes of thought with the double stochastic system of biological evolution. Is thought also characterized by such a double system? (If not, then the whole structure of this book is suspect.)

First it is important to note that what, in Chapter 1, I called "Platonism" is made possible today by arguments which are almost the opposite of those which a dualistic theology might prefer. The parallelism between biological evolution and mind is created not by postulating a Designer or Artificer hiding in the machinery of evolutionary process but, conversely, by postulating that thought is stochastic. The nineteenth-century critics of Darwin (especially Samuel Butler) wanted to introduce what they called "mind" (i.e., a supernatural entelechy) into the biosphere. Today I would emphasize that *creative* thought must always contain a random component. The exploratory processes—the endless *trial and error* of mental progress—can achieve the *new* only by embarking upon pathways randomly presented, some of which when tried are somehow selected for something like survival.

*Note that at a deep epistemological level, the *contrast* between the digital and the analogic is indeed a sharp contrast, such as occurs between components of digital systems. This contrast or discontinuity is a fundamental barrier between the somatic and the genetic (i.e., a barrier that prevents Lamarckian inheritance).

If we grant that creative thought is fundamentally stochastic, there are then several aspects of human mental process that suggest a positive analogy. We are looking for a binary division of thought process that will be stochastic in both of its halves, but the halves will differ in that the random component of one half will be digital and the random component of the other will be analogic.

The simplest way into this problem seems to be by considering first the selection processes that govern and limit the outcome. Here the two principal modes of testing thoughts or ideas are familiar.

The first is the test of coherence: Does the new idea make sense in terms of what is already known or believed? Granted that there are many sorts of sense and that "logic," as we have already seen, is a poor model of how the world operates, it is still so that some sort of consistency or coherence—rigorous or fanciful—is the thinker's first requirement of the notions which occur in the mind. Conversely, the genesis of new notions is almost totally (perhaps not totally) dependent upon reshuffling and recombining ideas that we already have.

There is, in fact, a remarkably close parallel between this stochastic process which goes on inside the brain and that other stochastic process which is the genesis of random genetic change on which an internal selection operates to ensure some conformity between the new and the old. And as we examine the matter more closely, the formal resemblance seems to increase.

In discussing the contrast between epigenesis and creative evolution, I pointed out that in epigenesis, all *new* information must be kept away and that the process is more like the elaborating of theorems within some primary tautology. I have pointed out in this chapter that the whole process of epigenesis can be viewed as an exact and critical filter, demanding certain standards of conformity within the growing individual.

We now note that in the intracranial process of thought, there is a similar filter that, like epigenesis within the individual organism, demands conformity and enforces this demand by a process more or less resembling logic (i.e., resembling the building up of tautology to create theorems). In the process of thought, *rigor* is the analogue of *internal coherence* in evolution.

In sum, the intracranial stochastic system of thought or learning closely resembles that component of evolution in which random genetic changes are selected by epigenesis. Finally, the cultural historian is provided with a world in which formal resemblances persist through many generations of cultural history, so that he can seek out such patterns just as a zoologist searches for homologies.

Turning now to that other process of learning or creative thought which involves not only the brain of the individual but also the world around the organism, we find the analogue of that process of evolution in which experience creates that relationship between creature and environment which we call *adaptation,* by enforcing changes of habit and soma.

Every action of the living creature involves some trial and error, and for any trial to be new, it must be in some degree random. Even if the new action is only a member of some well-explored *class* of actions, it must still, by its very newness, become in some measure a validation or exploration of the proposition "this is the way to do it."

But in learning, as in somatic change, there are limits and facilitations that select what can be learned. Some of these are external to the organism; others are internal. In the first instance, what can be learned at any given moment is limited or facilitated by what has previously been learned. In fact, there is a learning to learn with an ultimate limit, set by genetic constitution, to what can be immediately changed in response to environmental necessity. There is a peeling off, at each step, into genetic control (as noted in the discussion of somatic change in section 4).

Finally, it is necessary to put together the two stochastic processes which I have separated for the sake of analysis. What formal relation exists between the two?

As I see it, the root of the matter lies in the contrast between the digital and the analogic or, in another language, between the *name* and the *process* that is named.

But *naming* is itself a process and one that occurs not only in our analyses but profoundly and significantly within the systems we attempt to analyze. Whatever the coding and mechanical relation between DNA and the phenotype, DNA is still in some way a body of injunctions demanding—and in this sense, naming—the relations which shall become apparent in the phenotype.

And when we admit naming as a phenomenon occurring in and organizing the phenomena we study, we acknowledge ipso facto that in those phenomena, we expect hierarchies of logical typing.

So far we can go with Russell and *Principia.* But we are not now in Russell's world of abstract logic or mathematics and cannot accept an empty hierarchy of names or classes. For the mathematician, it is all very well to speak of *names of names of names* or of *classes of classes of classes.* But for the scientist, this empty world is insufficient. We are trying to deal with an interlocking or interaction of digital (i.e., naming) and analogic steps. *The process of naming is itself nameable,* and this fact compels us to substitute an *alternation* for the simple ladder of logical types that *Principia* would propose.

In other words, to recombine the two stochastic systems into which I have divided both evolution and mental process for the sake of analysis, I shall have to see the two as *alternating.* What in *Principia* appears as a ladder made of steps that are all alike (names of names of names and so on) will become an alternation of two species of steps. To get from the *name* to the *name of the name,* we must go through the *process* of naming the name. There must always be a generative process whereby the classes are created before they can be named.

This very large and complex matter will be the subject of Chapter 7.

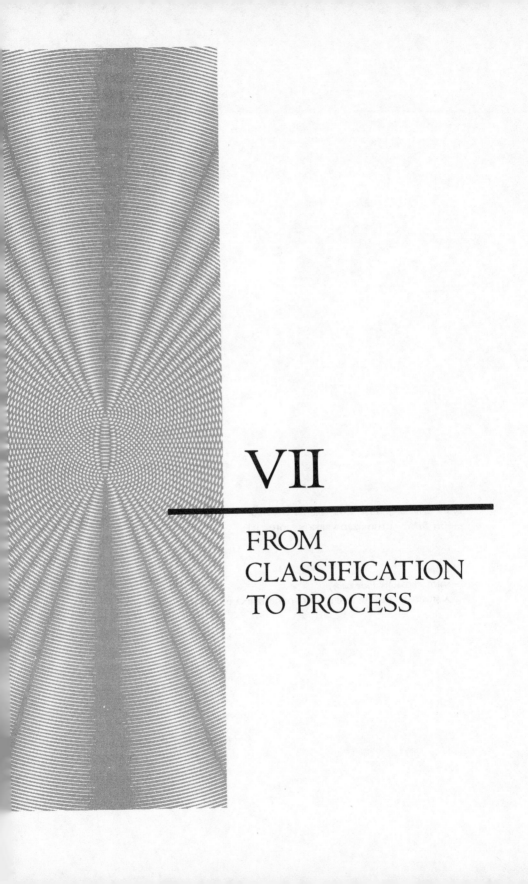

VII

FROM CLASSIFICATION TO PROCESS

In the beginning was the Word, and the Word was with God, and the Word was God.
—*Holy Bible,* AUTHORIZED VERSION, JOHN 1:1

Show me.
—SONG FROM *My Fair Lady,*
  A MUSICAL COMEDY BASED ON GEORGE BERNARD SHAW'S *Pygmalion.*

In Chapter 3, the reader was invited to contemplate a mixed batch of cases illustrating the near platitude that two descriptions are better than one. This series of cases ended with my description of what I regard as *explanation*. I asserted that at least one kind of explanation consists in supplementing the description of a process or set of phenomena with an abstract tautology onto which the description could be mapped. There may be other sorts of explanation, or it may be the case that all explanation in the end boils down to something like my definition.

It is surely the case that the brain contains no material objects other than its own channels and switchways and its own metabolic supplies and that all this material hardware never enters the narratives of the mind. Thought can be about pigs or coconuts, but there are no pigs

---

or coconuts in the brain; and in the mind, there are no neurons, only *ideas* of pigs and coconuts. There is, therefore, always a certain complementarity between the mind and the matters of its computation. The process of coding or representation that substitutes the idea of pigs or coconuts for the things is already a step, even a vast jump, in logical typing. The name is not the thing named, and the idea of pig is not the pig.

Even if we think of some larger circuit systems extending beyond the limits of the skin and call these systems *mind*, including within mind the man, his ax, the tree that he is felling, and the cut in the side of the tree;* even if all this be seen as a single system of circuits that meet the criteria of mind offered in Chapter 4; even so, there is no tree, no man, no ax in the mind. All these "objects" are only represented in the larger mind in the form of images and news of themselves. We may say that they propose themselves or propose their own characteristics.

In any case, it seems to me to be profoundly true that something like the relation which I have suggested between tautology and the matters to be explained obtains throughout the entire field of our inquiry. The very first step from pigs and coconuts into the world of coded versions plunges the thinker into an abstract and, I believe, a tautological universe. It is all very well to define explanation as "setting tautology and description side by side." This is only the beginning of the matter and would restrict explanation to the human species. Surely the dogs and cats, we might say, just accept things as they are, without all that ratiocination. But no. The thrust of my argument is that the very process of perception is an act of logical typing. Every image is a complex of many-leveled coding and mapping. And surely the dogs and cats have their visual images. When they look at you, surely they see "you." When a flea bites, surely the dog has an image of an "itch," located "there."

It still remains, of course, to apply this generalization to the realm of biological evolution. Before attempting that task, however, it is necessary to expand on the relationship between form and process, treating the notion of *form* as an analogue of what I have been calling *tautol-*

* See *Steps to an Ecology of Mind*, page 458.

ogy and *process* as the analogue of the aggregate of phenomena to be explained. As form is to process, so tautology is to description.

This dichotomy, which obtains in our own scientific minds as we look "out" upon a world of phenomena, is characteristic also of relationships among the very phenomena which we seek to analyze. The dichotomy exists on both sides of the fence between us and our subjects of discourse. The things-in-themselves (the *Dinge an sich*), which are inaccessible to direct inquiry, have relationships among themselves comparable to those relations that obtain between them and us. They, too (even those that are alive), can have no direct experience of each other—a matter of very great significance and a necessary first postulate for any understanding of the living world. What is crucial is the presupposition that ideas (in some very wide sense of that word) have a cogency and reality. They are what we can know, and we can know nothing else. The regularities or "laws" that bind ideas together—these are the "verities." These are as close as we can get to ultimate truth.

As a first step toward making this thesis intelligible, I will describe the process of my own analysis of a New Guinea culture.*

The work I had done in the field was shaped in no small degree by the arrival in New Guinea of a copy of the manuscript of Ruth Benedict's *Patterns of Culture* and by collaboration in the field with Margaret Mead and Reo Fortune. Margaret's theoretical conclusions from her fieldwork were published as *Sex and Temperament in Three Primitive Societies.*† The reader who is interested in dissecting out the story of the theoretical ideas in more detail is referred to my *Naven*, to Mead's *Sex and Temperament*, and of course, to Benedict's seminal *Patterns of Culture.***

Benedict had attempted to construct a typology of cultures using such terms as *Apollonian, Dionysian,* and *paranoid.* In *Sex and Temperament* and in *Naven*, the emphasis is shifted from characterization of cultural configurations to an attempt to characterize persons, the members of the cultures we had studied. We still used terms related to those which Benedict had used. Indeed, her typologies were borrowed from the lan-

---

* See Gregory Bateson, *Naven*, 1936. Reprint. Stanford, Calif.: Stanford University Press, 1958.
† New York: William Morrow & Co., 1935.
** New York: Houghton Mifflin & Co., 1934.

guage of the description of persons. I devoted a whole chapter of *Naven* to an attempt to use Kretschmer's old classification of persons into "cyclothyme"* and "schizothyme" temperaments. I treated this typology as an abstract map onto which I dissected my descriptions of Iatmul men and women.

This dissection and especially the fact of differentiating the typing of the sexes, which would have been foreign to the ideas of *Patterns of Culture,* led away from typology and into questions of process. It became natural to look at the Iatmul data as exemplifying those interactions between men and women which would create in the men and women that differentiation of ethos which was the base of my typology of persons. I looked to see how the behavior of the men might promote and determine that of the women, and vice versa.

In other words, I proceeded from a classification or typology to a study of the processes that generated the differences summarized in the typology.

But the next step was from process to a *typology of process.* I labeled the processes with the general term *schismogenesis,* and having put a label on the processes, I went on to a *classification* of them. It became clear that a fundamental dichotomy was possible. The processes of interaction that shared the general potentiality of promoting schismogenesis (i.e., first determining character within the individuals and beyond that creating intolerable stress) were, in fact, classifiable into two great genera: the symmetric and the complementary. I applied the term *symmetric* to all those forms of interaction that could be described in terms of competition, rivalry, mutual emulation, and so on (i.e., those in which A's action of a given kind would stimulate B to action of the same kind, which, in turn, would stimulate A to further similar actions. And so on. If A engaged in boasting, this would stimulate B to further boasting, and vice versa.)

In contrast, I applied the term *complementary* to interactional sequences in which the actions of A and B were different but mutually fitted each other (e.g., dominance-submission, exhibition-spectatorship,

---

* These almost obsolete terms were derived from the contrast between manic depressive and schizophrenic psychosis. *Cyclothyme* denoted the temperament of those who, according to Kretschmer, were prone to manic depressive psychosis, while *schizothyme* denoted the temperament of potential schizophrenics. See Kretschmer's *Physique and Character,* English translation 1925, and my *Naven,* 1936, Chapter 12.

dependence-nurturance). I noted that these paired relationships could likewise be schismogenic (e.g., that dependency might promote nurturance, and vice versa).

At this point, I had a classification or typology, not of persons, but of *processes,* and it was natural to swing from this classification to ask about what might be generated by interaction among the named processes. What would happen when symmetrical rivalry (which by itself would generate *symmetrical* schismogenesis of excessive competition) was mixed with *complementary* dependency-nurturance?

Sure enough, there were fascinating interactions between the named processes. It turned out that the symmetrical and complementary themes of interaction are mutually negating (i.e., have mutually opposite effects on relationship), so that when complementary schismogenesis (e.g., dominance-submission) has gone uncomfortably far, a little competition will relieve the strain; conversely, when competition has gone too far, a little dependency will be a comfort.

Later, under the rubric of *end-linkage,* * I investigated some of the possible permutations of combined complementary themes. It developed that a difference in premises, almost in choreography, between English and American middle-class cultures is related to the fact that spectatorship is preponderantly a filial function in England (i.e., is linked with dependency and submission) and preponderantly a parental function in America (i.e., is linked with nurturance and dominance).

That has all been spelled out elsewhere. What is important in the present context is to note that my procedures of inquiry were punctuated by an alternation between classification and the description of process. I had proceeded, without conscious planning, up an alternating ladder from description to the vocabulary of typology. But this typing of persons led back into a study of the processes by which the persons got that way. These processes were then classified into *types* of process types in their turn, were named by me. The next step was from the typing of process to study the interactions between the classified processes. This zigzag ladder between typology on one side and the study of process on the other is mapped in Figure 10.

* Bateson, G. "Regularities and Differences in National Character" in Watson, G., *Civilian Morale* (Boston: Houghton Mifflin, 1942). Reprinted in *Steps to an Ecology of Mind* (New York: Ballantine, 1972).

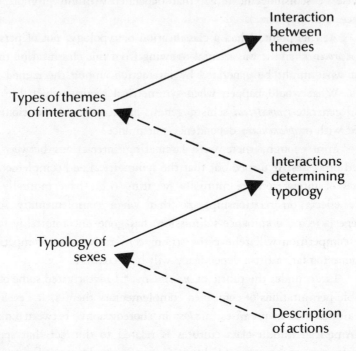

FORM PROCESS

Interaction
between
themes

Types of themes
of interaction

Interactions
determining
typology

Typology of
sexes

Description
of actions

Figure 10. Levels of analysis of Iatmul culture. The arrows mark the direction of my *argument.*

I shall now argue that the relations implicit or immanent in the events of the personal story I have just told (i.e., the zigzag sequence of steps from form to process and back to form) provide a very powerful paradigm for the mapping of many phenomena, some of which have already been mentioned.

I shall argue that this paradigm is not limited to a personal narrative of how a particular piece of theory came to be built, but that it recurs again and again wherever mental process as defined in Chapter 4 predominates in the organization of the phenomena. In other words, when we take the notion of logical typing out of the field of abstract logic and start to map real biological events onto the hierarchies of this paradigm, we shall immediately encounter the fact that in the world of mental and biological systems, the hierarchy is not only a list of classes, classes of classes, and classes of classes of classes but has also become a *zigzag ladder of dialectic between form and process.*

I shall further suggest that the very nature of perception follows this paradigm; that learning is to be modeled on the same sort of zigzag paradigm; that in the social world, the relation between love and marriage or education and status necessarily follow a similar paradigm; that in evolution, the relation between somatic and phylogenetic change and the relation between the random and the selected have this zigzag form. I shall suggest that similar relations obtain at a more abstract level between speciation and variation, between continuity and discontinuity and between number and quantity.

In other words, I am proposing that the relationship, which is rather ambiguously outlined in my story about analyzing a New Guinea culture, is, in fact, a relationship that will resolve a very large number of ancient puzzles and controversies in the fields of ethics, education, and evolutionary theory.

I begin from a discrimination I owe to Horst Mittelstaedt, who pointed out that there are two *sorts* of methods of perfecting an adaptive act.* Let us suppose that the act is the shooting of a bird. In the first case, this is to be done with a rifle. The marksman will look along the sights of his rifle and will note an error in its aim. He will correct that error, perhaps creating a new error which again he will correct, until he is satisfied. He will then press the trigger and shoot.

What is significant is that the act of self-correction occurs *within* the single act of shooting. Mittelstaedt uses the term *feedback* to characterize this whole genus of methods of perfecting an adaptive act.

In contrast, consider the case of the man who is shooting a flying bird with a shotgun or who uses a revolver held under the table where he cannot correct its aim. In such cases, what must happen is that an aggregate of information is taken in through sense organs; that upon this information, computation is completed; and that upon the (approximate) result of that computation, the gun is fired. There is no possibility of error correction in the single act. To achieve any improvement, correction must be performed upon a large *class* of actions. The man who would acquire skill with a shotgun or in the art of shooting pistols under the table must practice his art again and again, shooting at skeet or

---

* I owe the first step towards this insight to Mittelstaedt's presentation in 1960 of his study of how a praying mantis catches flies. See "The Analysis of Behavior in Terms of Control Systems" in *Transactions of the Fifth Conference on Group Process* (New York: Josiah Macy, Jr., Foundation, 1960).

some dummy target. By long practice, he must adjust the *setting* of his nerves and muscles so that in the critical event, he will "automatically" give an optimum performance. This genus of methods Mittelstaedt calls *calibration*.

It seems that, in these cases, "calibration" is related to "feedback" as higher logical type is related to lower. This relation is indicated by the fact that self-correction in the use of the shotgun is necessarily possible only from information derived from practice (i.e., from a *class* of past, completed actions).

It is, of course, true that skill in the use of the rifle can be increased by practice. The components of action that are so improved are common to the use of both rifle and shotgun. With practice, the marksman will improve his stance, learn to press the trigger without disturbing his aim, learn to synchronize his moment of firing with the moment of correcting his aim so that he does not overcorrect, and so on. These components of rifle shooting depend for improvement on practice and that calibration of nerve, muscle, and breathing which information from a class of completed actions will provide.

With respect to aim, however, the contrast of logical typing follows from the contrast between single instance and class of instances. It also appears that what Mittelstaedt calls *calibration* is a case of what I call *form* or *classification* and that his *feedback* is comparable to my *process*.

The next obvious question concerns the relation between the three dichotomies: form-process, calibration-feedback, and higher-lower logical type. Are these synonymous? I shall argue that form-process and calibration-feedback are indeed mutually synonymous but that the relation between higher and lower logical type is more complex. From what has already been said, it is clear both that structure may determine process and that, conversely, process may determine structure. It follows that there must be a relation between two levels of structure mediated by an intervening description of process. I believe that this is the analogue in the real world of Russell's abstract step from *class* to *class of classes*.

Let us consider the relation between feedback and calibration in a hierarchic example such as is provided by the temperature control in a dwelling house equipped with furnace, thermostat, and human resident (see Figure 11).

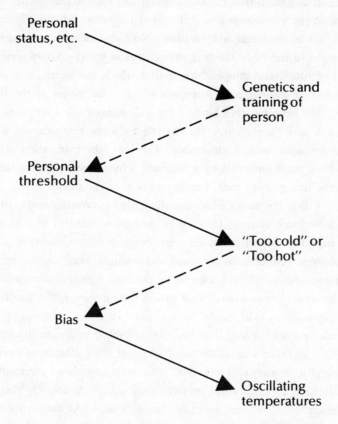

Personal
status, etc.

Genetics and
training of
person

Personal
threshold

"Too cold" or
"Too hot"

Bias

Oscillating
temperatures

Figure 11. Levels of Control of House Temperature. The arrows mark the direction of control.

At the lowest level, there is the temperature. This actual temperature from moment to moment (a *process*) affects a thermometer (a sort of sense organ) that is connected to the whole system in such a way that the temperature, as expressed by the bending of a double metal plate, will make or break an electric connection (a switch, a calibration) that controls the furnace. When the temperature rises above a certain point, the switch will be changed to the state called "OFF"; when the temperature falls below some lower point, the switch will be changed to "ON". The house will thus oscillate around some temperature between the two threshold points. At this level, the system is a simple, servo circuit such as I described in Chapter 4.

However, this simple feedback circuit is controlled by a calibration housed in the same small box that contains the thermometer. On the box is a knob that the householder can turn to change the setting, or *bias,* of the thermostat to a different temperature around which the temperature of the house will oscillate. Note that *two* calibrations have their location in the box: There is the control of state, ON/OFF, and the control of HIGH/LOW temperature around which the system will operate. If the former mean temperature was 65° F., the owner of the house may say, "It's been too cold lately." He will judge from a *sample* of his experiences and then change the setting to some temperature which will perhaps seem more comfortable. The bias (the calibration of the feedback) is itself governed by a feedback whose sense organ is located, not on the living room wall, but in the skin of the man.

But the man's bias—usually called his *threshold*—is, in turn, set by a feedback system. He may become more tolerant of cold as a result of hardship or exposure; he may become less tolerant as a result of prolonged residence in the tropics. He might even say to himself, "I'm getting too soft," and engage in outdoor training that will alter his calibration. Beyond that, what makes the man engage in special training or exposure to cold might be a change in status. He might become a monk or a soldier and thus become calibrated to a named social status.

In other words, the feedbacks and the calibrations alternate in a hierarchic sequence. Note that with each completed alternation (from calibration to calibration or from feedback to feedback), the sphere of relevance that we are analyzing has increased. At the simplest, lowest end of the zigzag ladder, the sphere of relevance was a furnace, ON or OFF; at the next level, a house oscillating around a certain temperature. At the next level, that temperature could be changed within a sphere of relevance that now included house *plus* resident over a much longer time, during which the man engaged in various outside activities.

With each zigzag of the ladder, the sphere of relevance increases. In other words, there is a change in logical typing of the information collected by the sense organ at each level.

Let us consider another example: A driver of an automobile travels at 70 miles per hour and thereby alerts the sense organ (radar, perhaps) of a traffic policeman. The bias or threshold of the policeman dictates that he shall respond to any difference greater than 10 miles per hour above or below the speed limit.

The policeman's bias was set by the local chief of police, who acted self-correctively with his eye on orders (i.e., calibration) received from the state capital.

The state capital acted self-correctively with the legislators' eyes on their voters. The voters, in turn, set a calibration within the legislature in favor of Democratic or Republican policy.

Again, we note an alternating ladder of calibration and feedback up to larger and larger spheres of relevance and more and more abstract information and wider decision.

Notice that within the system of police and law enforcement, and indeed in all hierarchies, it is most undesirable to have direct contact between levels that are nonconsecutive. It is not good for the total organization to have a pipeline of communication between the driver of the automobile and the state police chief. Such communication is bad for the morale of the police force. Nor is it desirable for the policeman to have direct access to the legislature, which would undermine the authority of the police chief.

To jump downward two or more steps in the hierarchy is likewise undesirable. The policeman should not have direct control over the accelerator or the brake system of the automobile.

The effect of any such jumping of levels, upward or downward, is that information appropriate as a basis for decision at one level will be used as basis for decision at some other level, a common variety of error in logical typing.

In legal and administrative systems, such jumping of logical levels is called *ex post facto* legislation. In families, the analogous errors are called *double binds*. In genetics, the Weissmannian barrier which prevents the inheritance of acquired characteristics seems to prevent disasters of this nature. To permit direct influence from somatic state to genetic structure might destroy the hierarchy of organization within the creature.

When we compare *learning* to shoot with a rifle with *learning* to shoot with a shotgun, another complication is introduced into the simple abstract paradigm of Russell's hierarchy of logical types. Both operations include cybernetic, self-corrective sequences. But the systemic difference between them is immediately evident when the sequences are viewed as contexts of learning.

The case of the rifle is comparatively simple. The error to be corrected (i.e., the information to be used) is the *difference* between the aim of the barrel and the direction of the target as disclosed by the alignment of sight and target. The marksman may have to go round and round this circuit many times, receiving news of error, correcting, receiving news of new error, correcting, receiving news of zero or minimal error, and firing.

But note that the marksman does not—or need not—carry forward news about what happened in the first round into his computation in the next round. The only relevant information is the error of the immediate moment. He does not need to change *himself*.

The man with a shotgun is in an entirely different position. For him, there is no separation between aim and firing that might allow him to correct his aim before he presses the trigger.* The aiming-and-firing, hyphenated, is a single act whose success or failure must be carried forward as information to the next act of firing. The entire operation must be improved, and therefore the entire operation is the subject matter of the information.

At the next act of shooting, the marksman must compute his action on the basis of the position of the new target *plus* information about what he did in the previous round of the cybernetic circuit *plus* information about the outcome of those actions.

In the third round of the circuit with another target, he should ideally use information about the *difference* between what happened in the first round and what happened in the second round. He might use the information at a nonverbal, kinesthetic level, saying to himself in muscular imagery, "that's what it felt like to overcorrect."

The rifleman simply goes round his cybernetic circuit a number of *separate* times; the man with a shotgun must accumulate his skill, packing his successive experiences, like Chinese boxes, each within the context of information derived from all previous relevant experiences.†

---

* I myself was taught to shoot during World War II, using an army automatic. The instructor had me stand with my back to a big tree and about six feet from it. My right hand had a grip on the weapon in its holster on my hip. I was to jump and turn as I jumped, raising the automatic and firing before my feet reached the ground. Preferably the bullet should hit the tree, but the speed and smoothness of the whole operation was more important than the accuracy.

† To ask about criteria of relevance would take us far afield into problems of contextual and other levels of learning.

From this paradigm, it appears that the idea of "logical typing," when transplanted from the abstract realms inhabited by mathematicological philosophers to the hurly-burly of organisms, takes on a very different appearance. Instead of a hierarchy of classes, we face a hierarchy of *orders of recursiveness*.

The question which I am asking of these instances of calibration and feedback concerns the necessity of differentiating between the two concepts in the real world. In the longer chains of description of house thermostat and law enforcement, is it so that the phenomena themselves contain (are characterized by) such a dichotomy of organization? Or is that dichotomy an artifact of my description? Can such chains be imagined *without* an immanent alternation of feedback and calibration? Is it perhaps so that such an alternation is basic to the way in which the world of adaptive action is put together? Should the characteristics of mental process (see Chapter 4) be extended to include terms of calibration and feedback?

There will surely be people who *prefer* to believe that the world is preponderantly punctuated by calibration, those typologists who, according to Ernst Mayr, can never understand natural selection. And there will be others who see only process or feedback.

Notably, Heraclitus, with his famous statement "into the same river no man can step twice," would be delighted by contemplation of the man with the shotgun. He might correctly say, "No man can shoot twice with a shotgun," because at every shooting, it will be a different man, differently calibrated. But later, remembering his dictum that everything flows; nothing is stationary, Heraclitus might turn around and deny the very existence of all calibration. After all, to be still is the essence of calibration. The still point is the setting of the turning world.

I believe that the resolution of this question depends upon our ideas of the nature of time (as also, the Russellian paradoxes of abstraction are resolved by the introduction of time into the argument; see Chapter 4).

The ongoing business of learning to shoot with a shotgun is *necessarily* discontinuous because the information about the self (i.e., the information required for calibration) can be harvested only *after* the moment of firing. Indeed, the firing of the gun is to the handling of it as the hen is to the egg. Samuel Butler's famous jest that the hen is an egg's way of making another egg should be corrected to say that the hen's later success in raising a family is the test of whether the egg from which she hatched was really a good egg. If the pheasant falls, the gun was well handled, the man well calibrated.

This view makes the process of learning to handle a gun necessarily discontinuous. The learning can occur only in separate increments at the successive moments of firing.

Similarly, the system of thermostatic control of the temperature of the house and the system of law enforcement are necessarily discontinuous for reasons connected with *time*. If any event is to depend upon some characteristic of a multiple sample of some other species of event, time must elapse for the accumulation of that sample, and this elapsed time will punctuate the dependent event to produce a discontinuity. But, of course, there would be no such "samples" in a world of purely physical causation. Samples are artifacts of description, creatures of mind, and shapers of mental process.

A world of sense, organization, and communication is not conceivable without discontinuity, without threshold. If sense organs can receive news only of difference, and if neurons either fire or or do not fire, then threshold becomes necessarily a feature of how the living and mental world is put together.

Chairoscuro is all very well, but William Blake tells us firmly that wise men see outlines and therefore they draw them.

VIII

SO
WHAT?

O, reason not the need: our basest beggars
Are in the poorest thing superfluous:
Allow not nature more than nature needs,
Man's life is cheap as beast's.

—SHAKESPEARE, *King Lear*

DAUGHTER: So what? You tell us about a few strong presuppositions and great stochastic systems. And from that we should go on to imagine how the world *is?* But—

FATHER: Oh, no. I also told you something about the limitations of imagining. So you should know that you cannot imagine the world as it *is*. (And why stress that little word?)

And I told you something about the self-validating power of ideas: that the world partly becomes—comes to be—how it is imagined.

DAUGHTER: Is that evolution, then? That going-on shifting and sliding of ideas to make all the ideas agree? But they never can.

FATHER: Yes, indeed. It all shifts and swirls around the verities. "Five plus seven will continue to equal twelve." In the world of ideas, numbers will still be in contrast with quantities. People will probably go on using *numerals* as names both for quantities and for numbers. And they'll go on being misled by their own bad habits. And so on. But, yes, your image of evolution is exact. And what Darwin called "natural selection" is the surfacing of the tautology or presupposition that what stays true longer does indeed stay true longer than what stays true not so long.

DAUGHTER: Yes, I know you love reciting that sentence. But do the verities stay true forever? And are these things you call *verities* all tautological?

FATHER: Wait, wait. There are at least three questions all tied together. Please.

First, *no*. Our opinions about the verities are surely liable to change.

Second, whether the verities that Saint Augustine called *eternal* verities are true forever apart from our opinions, I cannot know.

DAUGHTER: But can you *know* if it's all tautological?

FATHER: No, of course not. But if the question is once asked, I cannot avoid having an opinion.

DAUGHTER: Well, is it?

FATHER: Is it what?

DAUGHTER: Tautological?

FATHER: All right. My opinion is that the Creatura, the world of mental process, is both tautological and ecological. I mean that it is a slowly self-healing tautology. Left to itself, any large piece of Creatura will tend to settle toward tautology, that is, toward *internal consistency* of ideas and processes. But every now and then, the consistency gets torn; the tautology breaks up like the surface of a pond when a stone is thrown into it. Then the tautology slowly but immediately starts to heal. And the healing may be ruthless. Whole species may be exterminated in the process.

DAUGHTER: But, Daddy, you could make consistency out of the idea that it always starts to heal.

FATHER: So, the tautology is not broken; it's only pushed up to the next level of abstraction, the next logical type. That's so.

DAUGHTER: But how many levels are there?

FATHER: No, that I cannot know. I cannot know whether it is ultimately a tautology nor how many logical levels it has. I am inside it and therefore cannot know its outer limits—if it has any.

DAUGHTER: I think it's gloomy. What's the point of it all?

FATHER: No, no. If you were in love, you would not ask that question.

DAUGHTER: You mean that love is the point?

FATHER: But again, no. I was saying no to your question, not answering it. It's a question for an occidental industrialist and an engineer. This whole book is about the wrongness of that question.

DAUGHTER: You never said that in the book.

FATHER: There are a million things I never said. But I'll answer your question. It has a million—an infinite number—of "points," as you call them.

DAUGHTER: But that's like having no point—Daddy, is it a sphere?

FATHER: Ah, all right. That will do for a metaphor. A multidimensional sphere, perhaps.

DAUGHTER: Hmm—a self-healing tautology, which is also a sphere, a multidimensional sphere.

DAUGHTER: So what?

FATHER: But I keep telling you: There is no "what." A million points or none.

DAUGHTER: Then why write this book?

FATHER: That's different. This book, or you and me talking, and so on—these are only little pieces of the bigger universe. The total self-healing tautology has no "points" that you can enumerate. But when you break it up into little pieces, that's another story. "Purpose" appears as the universe is dissected. What Paley called "design" and Darwin called "adaptation."

DAUGHTER: Just an artifact of dissection? But what's dissection for? This whole book is a dissection. What's it for?

FATHER: Yes, it's partly dissection and partly synthesis. And I suppose that under a big enough macroscope, no idea can be wrong, no purpose destructive, no dissection misleading.

DAUGHTER: You said that we only *make* the parts of any whole.

FATHER: No, I said that parts are *useful* when we want to describe wholes.

DAUGHTER: So you want to describe wholes? But when you've done it, what then?

FATHER: All right, let's say we live, as I said, in a self-healing tautology that is more or less often getting torn more or less badly. That seems to be how it is in our neighborhood of space-time. I guess some tearing of the tautological ecological system is even—in a way—good for it. Its capacity for self-healing may need to be exercized, as Tennyson says, "lest one good custom should corrupt the world."

And, of course, death has that positive side. However good the man, he becomes a toxic nuisance if he stays around too long. The blackboard, where all the information accumulates, must be wiped off, and the pretty lettering on it must be reduced to random chalky dust.

DAUGHTER: But—

FATHER: And so on. There are subcycles of living and dying within the bigger, more enduring ecology. But what shall we say of the death of the larger system? *Our* biosphere? Perhaps under the eye of heaven or Shiva, it doesn't matter. But it's the only one we know.

DAUGHTER: But your book is a part of it.

FATHER: Of course it is. But, yes, I see what you mean, and of course you are right. Neither the deer nor the mountain lion needs an excuse for being, and my book, too, as part of the biosphere, needs no excuses. Even if I'm all wrong!

DAUGHTER: Can the deer or the mountain lion be *wrong?*

FATHER: Any species can get into an evolutionary cul-de-sac, and I suppose it is a mistake of sorts for that species to be a party to its own extinction. The human species, as we all know, may extinguish itself any day now.

DAUGHTER: So what? Why write the book?

FATHER: And there is some pride in it, too, a feeling that if we are all going down to the sea like lemmings, there should be at least one lemming taking notes and saying, "I told you so." To believe that I could stop the race to the ocean would be even more arrogant than saying, "I told you so."

DAUGHTER: I think you are talking nonsense, Daddy. I don't see you as the only intelligent lemming taking notes on the self-destruction of the others. It's not like you—so there. Nobody is going to buy a book by a sardonic lemming.

FATHER: Oh, yes. It's nice to have a book sell, but always a surprise, I guess. Anyhow that's not what we are talking about. (And you'd be surprised at how many books by sardonic lemmings do, in fact, sell very nicely.)

DAUGHTER: So what?

FATHER: For me, after fifty years of pushing these ideas about, it has slowly become clear that muddleheadedness is not necessary. I have always hated muddleheadedness and always thought it was a necessary condition for religion. But it seems that that is not so.

DAUGHTER: Oh, is *that* what the book is about?

FATHER: You see, they preach *faith,* and they preach *surrender.* But I wanted *clarity.* You could say that faith and surrender were necessary to maintain the search for clarity. But I have tried to avoid the sort of faith that would cover up the gaps in the clarity.

DAUGHTER: Go on.

FATHER: Well, there were turning points. One of them was when I saw that the Fraserian idea of magic was upside down or inside out. You know, the conventional view is that religion evolved out of magic, but I think it was the other way around—that magic is a sort of degenerate religion.

DAUGHTER: So what do you *not* believe?

FATHER: Well, for example, I do not believe that the original purpose of the rain dance was to make "it" rain. I suspect that that is a degenerate misunderstanding of a much more profound religious need: to affirm membership in what we may call the *ecological tautology,* the eternal verities of life and environment.

There's always a tendency—almost a need—to vulgarize religion, to turn it into entertainment or politics or magic or "power."

DAUGHTER: And ESP? And materialization? And out-of-body experience? And spiritualism?

FATHER: All symptoms, mistaken attempts at cute efforts to escape from a crude materialism that becomes intolerable. A miracle is a materialist's idea of how to escape from his materialism.

DAUGHTER: Is there no escape? I don't understand.

FATHER: Oh, yes. But, you see, magic is really only a sort of pseudoscience. And like applied science, it always proposes the possibility of *control*. So you don't get away from all that way of thought by sequences into which that way of thinking is already built-in.

DAUGHTER: So how do you get away?

FATHER: Ah, yes. The reply to crude materialism is not miracles but beauty—or, of course, ugliness. A small piece of Beethoven symphony, a single Goldberg variation, a single organism, a cat or a cactus, the twenty-ninth sonnet or the Ancient Mariner's sea snakes. You remember he "blessed them, unaware," and the Albatross then fell from his neck into the sea.

DAUGHTER: But you didn't write that book. That's the one you should have written. The one about the Albatross and the Symphony.

FATHER: Ah, yes. But, you see, I couldn't do that. This book had to be done first. Now, after all the discussion of mind and tautology and immanent differences and so on, I am beginning to be ready for symphonies and albatrosses. . . .

DAUGHTER: Go on.

FATHER: No, you see it's not possible to map beauty-and-ugliness onto a flat piece of paper. Oh yes, a drawing may be beautiful and on flat paper but that's not what I'm talking about. The question is onto what surface shall a *theory* of aesthetics be mapped? If you ask me that question today I could attempt an answer. But not two years ago when this book was still unwritten.

DAUGHTER: All right. So today how would you answer?

FATHER: And then there's *consciousness* which I have not touched—or touched only once or twice—in this book. Consciousness and aesthetics are the great untouched questions.

DAUGHTER: But whole rooms in libraries are full of books about those "untouched" questions.

FATHER: No, no. What is untouched is the question: Onto what sort of surface shall "aesthetics" and "consciousness" be mapped?

DAUGHTER: I don't understand.

FATHER: I mean something like this: That both "consciousness" and "aesthetics" (whatever those words mean) are either characteristics present in all *minds* (as defined in this book), or they are spinoffs—late fancy creations from such minds. In either case, it is the primary definition of mind that has to accommodate the theories of aesthetics and consciousness. It's onto that primary definition that the next step must be mapped. The terminology to deal with beauty-ugliness and the terminology for consciousness have got to be elaborated out of (or mapped onto) the ideas in the present book or similar ideas. It's that simple.

DAUGHTER: Simple?

FATHER: Yes. Simple. I mean the proposition that that is what must be done is simple and clear. I don't mean that the *doing* will be simple.

DAUGHTER: Well. How would you begin?

FATHER: *Il n'y a que le premier pas qui coûte.* It's the first step that is difficult.

DAUGHTER: All right. Never mind about that. *Where* would you begin?

FATHER: There has to be a reason why these questions have never been answered. I mean, we might take that as our first clue to the answer—the historical fact that so many men have tried and not succeeded. The answer must be somehow hidden. It must be so: That the very posing of these questions always gives a false scent, leading the questioner off on a wild goose chase. A red herring.

DAUGHTER: Well?

FATHER: So let's look at the "schoolboy" truisms that I have put together in this book to see if one or more of those could hide answers to the questions of consciousness or aesthetics. I'm sure that a person or a poem or a pot . . . or a landscape . . .

DAUGHTER: Why don't you make a list of what you call the "schoolboy" points? Then we could try the ideas, "consciousness" and "beauty" on the list.

FATHER: Here is a list. First there were the six criteria of *mind* (Chapter 4):

1. Made of parts which are not themselves mental. "Mind" is immanent in certain sorts of *organization* of parts.

2. The parts are triggered by events in time. Differences though static in the outside world can generate events if *you* move in relation to them.

3. Collateral energy. The stimulus (being a difference) may provide no energy but the respondent has energy, usually provided by metabolism.

4. Then causes-and-effects form into circular (or more complex) chains.

5. All messages are coded.

6. And last, most important, there is the fact of logical typing.

Those are all fairly well-defined points and they support each other pretty well. Perhaps the list is redundant and could be reduced, but that's not important at this moment. Beyond those five points, there is the remainder of the book. And that is about different sorts of what I called *double description* and ranging from binocular vision to the combined effect of the "great" stochastic processes and the combined effect of "calibration" and "feedback." Or call it "rigor and imagination" or "thought and action."

That's all.

DAUGHTER: All right. So where would you attach the phenomena of beauty and ugliness and consciousness?

FATHER: And don't forget the *sacred*. That's another matter that was not dealt with in the book.

DAUGHTER: Please, Daddy. Don't do that. When we get near to asking a question, you jump away from it. There's always another question it seems. If you could answer *one* question. Just one.

FATHER: No. You don't understand. What does e.e. cummings say? "Always the more beautiful answer who asks the more difficult question." Something like that. You see I am not asking another question each time. I am making the same question bigger. The *sacred* (whatever that means) is surely related (somehow) to the *beautiful* (whatever that means). And if we could say how they are related, we could perhaps say what the words mean. Or perhaps that would never be necessary. Every time we add a related piece to the question, we get more clues to what sort of answer we should expect.

DAUGHTER: So now we have six pieces of the question?

FATHER: Six?

DAUGHTER: Yes. It was two at the beginning of this conversation. Now it's six. There's consciousness, and beauty and the sacred, and then there's the relation between consciousness and beauty, and the relation between beauty and the sacred, and the relation between the sacred and consciousness. That makes six.

FATHER: No. Seven. You're forgetting the book. All your six make up together a triangular sort of question and that triangle is to be related to what's in this book.

DAUGHTER: All right. Go on. Please.

FATHER: I think I would like to call my next book *"Where Angels Fear to Tread."* Everybody keeps wanting me to rush in. It is monstrous—vulgar, reductionist, sacreligious—call it what you will—to rush in with an over-simplified question. It's a sin against all three of our new principles. Against aesthetics and against consciousness and against the sacred.

DAUGHTER: But where?

FATHER: Ah. Yes. That question proves the close relationship between consciousness and beauty and the sacred. It is consciousness running around like a dog with its tongue out—literally cynicism—

that asks the too simple question and shapes up the vulgar answer. To be conscious of the nature of the sacred or of the nature of beauty is the folly of reductionism.

DAUGHTER: Is all that related to this book?

FATHER: Yes. Yes indeed it is. Chapter 4, the listing of the criteria, if it stood alone, would be "gross," as the kids say. A vulgar answer to an oversimplified question. Or an oversimplified answer to a vulgar question. But, precisely the elaboration of discussion about "double description," "structure and process," and double stochastic systems—that elaboration saves the book from vulgarity. I hope so at least.

DAUGHTER: And the next book?

FATHER: Will start from a map of the region *where angels fear to tread.*

DAUGHTER: A vulgar map?

FATHER: Perhaps. But I do not know what will follow the map and enclose it in some wider and more difficult question.

APPENDIX:
TIME
IS OUT
OF JOINT*

* A memorandum circulated to the Regents of
the University of California, August 1978.

At the meeting of the Committee on Educational Policy, July 20, 1978, I remarked that current educational processes are a *"rip off,"* from the point of view of the student. The present note is to explain this view.

It is a matter of *obsolescence*. While much that universities teach today is new and up to date, the presupposition or premises of thought upon which all our teaching is based are ancient and, I assert, *obsolete*.

I refer to such notions as:

a. The Cartesian dualism separating "mind" and "matter."

b. The strange physicalism of the metaphors which we use to describe and explain mental phenomena—"power," "tension," "energy," "social forces," etc.

c. Our anti-aesthetic assumption, borrowed from the emphasis which Bacon, Locke, and Newton long ago gave to the physical sciences, viz. that all phenomena (including the mental) can and shall be studied and *evaluated* in quantitative terms.

The view of the world—the latent and partly *unconscious* epistemology—which such ideas together generate is out of date in three different ways:

a. Pragmatically, it is clear that these premises and their corollaries lead to greed, monstrous over-growth, war, tyranny, and pollution. In this sense, *our* premises are daily demonstrated false, and the students are half aware of this.

b. *Intellectually,* the premises are obsolete in that systems theory, cybernetics, holistic medicine, ecology, and gestalt psychology offer demonstrably better ways of understanding the world of biology and behaviour.

c. As a base for *religion,* such premises as I have mentioned became *clearly intolerable and therefore obsolete* about 100 years ago. In the aftermath of Darwinian evolution, this was stated rather clearly by such thinkers as Samuel Butler and Prince Krapotkin. But already in the eighteenth century, William Blake saw that the philosophy of Locke and Newton could only generate "dark Satanic mills."

Necessarily every aspect of our civilization is split wide open. In the field of economics, we face two overdrawn caricatures of life—the capitalist or the communist—and we are told that we *must* take sides in the struggle between these two monstrous ideologies. In the business of thinking, we are torn between various extremes of affectlessness and the strong current of anti-intellectual fanaticism.

As in religion, the constitutional guarantees of "religious freedom" seem to promote similar exaggerations: a strange, totally secular Protestantism, a wide spectrum of magical cults, and total religious ignorance. It is no accident that simultaneously the Roman Catholic Church is giving up the use of Latin, while the rising generation is learning to chant in Sanskrit!

So, in this world of 1978, we try to run a university and to maintain standards of "excellence" in the face of growing *distrust, vulgarity, insanity, exploitation of resources, victimisation of persons,* and *quick commercialism.* The screaming voices of greed, frustration, fear, and hate.

It is understandable that the Board of Regents concentrates attention upon matters which can be handled at a superficial level, avoiding the swamps of all sorts of extremism. But I still think that the facts of deep obsolescence will, in the end, compel attention.

As a technical school, we do pretty well. We can at least teach young people to be engineers, doctors, lawyers. We can confer the skills which lead to success in trades whose working philosophy is again the same old dualistic pragmatism. And that is much. It is perhaps not the main duty and function of a great university. . . .

But do not get the idea that the faculty and the administration and the regents are the only obsoletes, while the students are wise and noble and up-to-date. *They are just as obsolete as we.* We are all in the same boat, whose name is "ONLY 1978," the time which is out of joint. In 1979 we shall know a little more by dint of rigor and imagination, the two great contraries of mental process, either of which by itself is lethal. Rigor alone is paralytic death, but imagination alone is insanity.

Tweedledum and Tweedledee *agreed* to have a battle; and isn't it a blessing that the contrasting generations can agree that social "power" has physical dimensions and can engage in battles for this strange abstraction. (In other times and other places, battles were fought for "honor," "beauty," and even "truth." . . .)

Looking at the whole mess from another angle, I believe that the students were right in the sixties: There was something very wrong in their education and indeed in almost the whole culture. But I believe that they were wrong in their diagnosis of where the trouble lay. They fought for "representation" and "power." On the whole, they won their battles and now we have student representation on the Board of Regents and elsewhere. But it becomes increasingly clear that the winning of these battles for "power" has made no difference in the educational process. The obsolescence to which I referred is unchanged and, no doubt, in a few years we shall see the same battles, fought over the same phoney issues, all over again.

There really is something deeply wrong . . . and I am not convinced that what is wrong is a necessary tribulation about which nothing can be done.

A sort of freedom comes from recognizing what is necessarily so. After that is recognized, comes a knowledge of how to act. You can ride a bicycle only after your partly unconscious reflexes acknowledge the laws of its moving equilibrium.

I must now ask you to do some thinking more technical and more theoretical than is usually demanded of general boards in their perception of their own place in history. I see no reason why the regents of a great university should share in the anti-intellectual preferences of the press or media. Indeed to force these preferences upon them would be insulting.

I therefore propose to analyze the lopsided process called "obsolescence" which we might more precisely call "one-sided progress." Clearly for obsolescence to occur there must be, in other parts of the system, other changes compared with which the obsolete is somehow lagging or left behind. In a static system, there would be no obsolescence!

It seems that there are two components in evolutionary process, and that mental process similarly has a double structure. Let me use biological evolution as a parable or paradigm to introduce what I want to say later about thought, cultural change and education.

Survival* depends upon two contrasting phenomena or processes, two ways of achieving adaptive action. Evolution must always, Janus-like, face in two directions: inward towards the developmental regularities and physiology of the living creature and outward towards the vagaries and demands of the environment. These two necessary components of life contrast in interesting ways: the inner development—the embryology or "epigenesis"—is *conservative* and demands that every new thing shall conform or be compatible with the regularities of the *status quo ante*. If we think of a natural selection of new features of anatomy or physiology—then it is clear that one side of this selection process will favor those new items which do not upset the old apple cart. This is minimal necessary conservatism.

In contrast, the outside world is perpetually changing and becoming ready to receive creatures which have undergone change, almost insisting upon change. No animal or plant can ever be "ready made." The internal recipe insists upon compatibility but is never sufficient for the development and life of the organism. Always the creature itself must achieve change of its own body. It must acquire certain somatic characteristics by use, by disuse, by habit, by hardship, and by nurture. These *"acquired characteristics"* must, however, never be passed on to the offspring. They must not be directly incorporated into the DNA. In organisational terms, the injunction—e.g., to make babies with strong shoulders who will work better in coal mines—must be transmitted *through channels,* and the channel in this case is *via* natural external selection of those offspring who happen (thanks to the *random* shuffling of genes and random creation of mutations) to have a greater propensity for developing stronger shoulders under the stress of working in coal mines.

The individual body undergoes adaptive change under external pressure, but natural selection acts upon the gene pool of the *population.* But note this principle which biologists commonly overlook, that it is an acquired characteristic called *"working in coal mines"* which sets the context for the selection of the genetic changes called "increased propensity for developing stronger shoulders." The acquired characteristics do not become unimportant by not being carried in and passed on by DNA. It is still *habits* which set the conditions for natural selection.

And note this converse principle that the acquisition of bad habits, at a

* By survival, I mean the maintenance of a steady state through successive generations. Or, in negative terms, I mean the avoidance of the *death of the largest system about which we can care*. Extinction of the dinosaurs was trivial in galactic terms but this is no comfort to them. We cannot care much about the inevitable survival of systems larger than our own ecology.

social level, surely sets the context for selection of ultimately lethal genetic propensities.

We are now ready to look at obsolescence in mental and cultural processes.

If you want to understand mental process, look at biological evolution and conversely if you want to understand biological evolution, go look at mental process.

I called attention above to the circumstance that internal selection in biology must always stress *compatibility* with the immediate past and that over long evolutionary time it is internal selection which determines those "homologies" which used to delight a previous generation of biologists. It is internal selection which is conservative and this conservatism shows itself most strongly in embryology and in the preservation of abstract form.

The familiar mental process by which a tautology* grows and differentiates into multiple theorems resembles the process of embryology.

In a word, conservatism is rooted in *coherence* and *compatibility* and these go along with what, above, I called *rigor* in the mental process. It is here that we must look for the roots of obsolescences.

And the paradox or dilemma which perplexes and dismays us when we contemplate correcting or fighting against obsolescence is simply the fear that we must lose coherence and clarity and compatibility and *even sanity,* if we let go of the obsolete.

There is however another side to obsolescence. Clearly if some part of a cultural system "lags behind," there must be some other part which has evolved "too fast." Obsolescence is in the contrast between the two components. If the lagging of one part is due to the internal half of natural selection, then it is natural to guess that the roots of too rapid "progress"—if you please—will be found in the processes of external selection.

And, sure enough, that is precisely what is the case. "Time is out of joint" because these two components of the steering of evolutionary process are mutually out of step: Imagination has gone too far ahead of rigor and the result looks, to conservative elderly persons like me, remarkably like insanity or perhaps like nightmare, the sister of insanity. Dream is a process, uncorrected by either internal rigor or external "reality."

In certain fields, what I have said above is already familiar. Notoriously the law lags behind technology, and notoriously the obsolescence which goes

---

* "Tautology" is the technical term for such aggregates or networks of propositions as Euclidean geometry, Riemannian geometry, or arithmetic. The aggregate springs from a set cluster of arbitrary axioms or definitions and *no "new" information may be added* to that cluster after the assertion of axioms. The "proof" of a theorem is the demonstration that indeed the theorem was entirely latent in the axioms and definitions.

with senescence is an obsolescence of ways of thought which makes it difficult for the old to keep up with the mores of the young. And so on.

But I have said a little more than these particular examples could convey. It seems that these are examples of a very profound and general principle, whose wide generality is demonstrated by its being applicable to evolutionary as well as to mental process.

We are dealing with a species of abstract relation which recurs as a necessary component in many processes of change and which has many names. Some of its names are familiar: pattern/quantity, form/function, letter/spirit, rigor/imagination, homology/analogy, calibration/feedback, and so on.

Individual persons may favor one or the other component of this dualism and we will then call them "conservatives," "radicals," "liberals," and so on. But behind these epithets lies epistemological truth which will insist that the poles of contrast dividing the persons are indeed dialectical necessities of the living world. You cannot have "day" without "night," nor "form" without "function."

The practical problem is of combination. How, recognizing the dialectic relation between these poles of contrast, shall we proceed? To play one half of the adversarial game would be easy, but *statesmanship* requires something more and, truly, more difficult.

I suggest that if the Board of Regents has any non-trivial duty it is that of statesmanship in precisely this sense—the duty of rising above partisanship with any component or particular fad in university politics.

Let us look at how the contrasts between form and function, etc. are met, remembering that the problem is always a matter of timing: How shall change in form be *safely* speeded up to avoid obsolescence? And how shall descriptions of change in functioning be summarized and coded, not too fast, into the corpus of form?

The rule in biological evolution is plain: The immediate individual bodily effects of functioning shall never be allowed to impinge upon the individual genetic coding. The gene pool of the *population* is however subject to change under a natural selection which will recognize differences, especially differences in ability to achieve more adaptive functioning. The barrier which prohibits "Lamarckian" inheritance precisely protects the gene system from too rapid change under possibly capricious environmental demands.

But in cultures and social systems and great universities there is no equivalent barrier. Innovations become irreversibly adopted into the on-going system without being tested for long-time viability; and necessary changes are resisted by the core of conservative individuals without any assurance that these particular changes are the ones to resist.

Individual comfort and discomfort become the only criteria for choice of *social* change and the basic contrast of logical typing between the member and the category is forgotten until new discomforts are (inevitably) created by the new state of affairs. Fear of individual death and grief propose that it would be "good" to eliminate epidemic disease and only after 100 years of preventive medicine do we discover that the population is overgrown. And so on.

Obsolescence is not to be avoided by simply speeding up change in structure, nor can it be avoided by simply slowing down functional changes. It is clear that neither an over-all conservatism nor an over-all eagerness for change is appropriate. An adversarial combination of the two habits of mind would perhaps be better than either habit alone but, adversarial systems are notoriously subject to irrelevant determinism. The relative "strength" of the adversaries is likely to rule the decision regardless of the relative strength of their arguments.

It is not so much "power" that corrupts as the *myth* of "power." It was noted above that "power," like "energy," "tension," and the rest of the quasi-physical metaphors are to be distrusted and, among them, "power" is one of the most dangerous. He who covets a mythical abstraction must always be insatiable! As teachers we should not promote that myth.

It is difficult for an adversary to see further than the dichotomy between winning and losing in the adversarial combat. Like a chess player, he is always tempted to make a tricky move, to get a quick victory. The discipline, always to look for the best move on the board, is hard to attain and hard to maintain. The player must have his eye always on a longer view, a larger gestalt.

So we come back to the place from which we started—seeing that place in a wider perspective. The place is a university and we its Board of Regents. The wider perspective is *about* perspectives, and the question posed is: Do we, as a board, foster whatever will promote in students, in faculty, and around the boardroom table those wider perspectives which will bring our system back into an appropriate synchrony or harmony between rigor and imagination?

As *teachers,* are we wise?

G.B.

GLOSSARY

Adaptation. A feature of an organism whereby it seemingly fits better into its
  environment and way of life. The process of achieving that fit.

Analogic. See *Digital.*

Brownian movement. The constant movement of molecules, zigzag and unpredictable, caused by their mutual impacts.

Co-Evolution. A stochastic system of evolutionary change in which two or more
  species interact in such a way that changes in species A set the stage for
  the natural selection of changes in species B. Later changes in species B,
  in turn, set the stage for the selecting of more similar changes in species
  A.

Cybernetics. A branch of mathematics dealing with problems of control, recursiveness, and information.

Digital. A signal is *digital* if there is discontinuity between it and alternative
  signals from which it must be distinguished. *Yes* and *no* are examples of
  digital signals. In contrast, when a magnitude or quantity in the signal

is used to represent a continuously variable quantity in the referent, the signal is said to be *analogic*.

Eidetic. A mental image is *eidetic* if it has all the characteristics of a percept, especially if it is referred to a sense organ and so seems to come in from the outside.

Energy. In this book, I use the word *energy* to mean a *quantity* having the dimensions: mass times velocity squared (MV^2). Other people, including physicists, use it in many other senses.

Entropy. The degree to which relations between the components of any aggregate are mixed up, unsorted, undifferentiated, unpredictable, and random (q.v.). The opposite is *negentropy*, the degree of ordering or sorting or predictability in an aggregate. In physics, certain sorts of ordering are related to quantity of available energy.

Epigenesis. The processes of embryology seen as related, at each stage, to the status quo ante.

Epistemology. A branch of science combined with a branch of philosophy. As science, epistemology is the study of how particular organisms or aggregates of organisms *know, think,* and *decide.* As philosophy, epistemology is the study of the necessary limits and other characteristics of the processes of knowing, thinking, and deciding.

Flexibility. See *Stress.*

Genetic. Strictly, the science of genetics deals with all aspects of the heredity and variation of organisms and with the processes of growth and differentiation within the organism.

Genotype. The aggregate of recipes and injunctions that are the hereditary contributions to the determination of the phenotype (q.v.).

Homology. A formal resemblance between two organisms such that the relations between certain parts of A are similar to the relations between corresponding parts of B. Such formal resemblance is considered to be evidence of evolutionary relatedness.

Idea. In the epistemology offered in this book, the smallest unit of mental process is a difference or distinction or news of a difference. What is called an *idea* in popular speech seems to be a complex aggregate of such units. But popular speech will hesitate to call, say, the bilateral symmetry of a frog or the message of a single neural impulse an *idea*.

Information. Any difference that makes a difference.

Linear and *lineal.* *Linear* is a technical term in mathematics describing a relationship between variables such that when they are plotted against each other on orthogonal Cartesian coordinates, the result will be a straight line. *Lineal* describes a relation among a series of causes or arguments such that the sequence does not come back to the starting point. The opposite of *linear* is *nonlinear.* The opposite of *lineal* is *recursive.*

Logical types. A series of examples is in order:

1. The name is not the thing named but is of different logical type, higher than that of the thing named.

2. The class is of different logical type, higher than that of its members.

3. The injunctions issued by, or control emanating from, the bias of the house thermostat is of higher logical type than the control issued by the thermometer. (The *bias* is the device on the wall that can be set to determine the temperature around which the temperature of the house will vary.)

4. The word *tumbleweed* is of the same logical type as *bush* or *tree*. It is not the name of a species or genus of plants; rather, it is the name of a class of plants whose members share a particular style of growth and dissemination.

5. *Acceleration* is of higher logical type than *velocity*.

Mutation. In conventional evolutionary theory, offspring may differ from their parents for the following sorts of reasons:

1. Changes in DNA called *mutations*.

2. Reshuffling of genes in sexual reproduction.

3. Somatic changes acquired during the individual's life in response to environmental pressure, habit, age, and so forth.

4. Somatic segregation, that is, the dropping or reshuffling of genes in epigenesis resulting in patches of tissue that have differentiated genetic makeup. Genetic changes are always digital (q.v.), but modern theory prefers (with good reason) to believe that *small* changes are, in general, the stuff of which evolution is made. It is assumed that many small mutational changes combine over many generations to make larger evolutionary contrasts.

Negentropy. See *Entropy*.

Ontogeny. The process of development of the individual; embryology *plus* whatever changes environment and habit may impose.

Parallax. The *appearance* of movement in observed objects, which is created when the observer's eye moves relative to them; the difference between the apparent positions of objects seen with one eye and their apparent positions as seen with the other eye.

Phenocopy. A phenotype (q.v.) that shares certain characteristics with other phenotypes in which these characteristics are brought about by genetic factors. In the *phenocopy*, these characteristics are brought about by somatic change under environmental pressure.

Phenotype. The aggregate of propositions making up the description of a real organism; the appearance and characteristics of a real organism. See *Genotype*.

Phylogeny. The evolutionary history of a species.

Prochronism. The general truth that organisms carry, in their forms, evidences of their past growth. Prochronism is to ontogeny as homology (q.v.) is to phylogeny.

Random. A sequence of events is said to be *random* if there is no way of predicting the next event of a given kind from the event or events that have preceded and if the system obeys the regularities of probability. Note that the events which we say are *random* are always members of some limited set. The fall of an honest coin is said to be *random*. At each throw, the probability of the next fall being heads or tails remains unchanged. But the randomness is within the limited set. It is heads or tails; no alternatives are to be considered.

Reductionism. It is the task of every scientist to find the simplest, most economical, and (usually) most elegant explanation that will cover the known data. Beyond this, reductionism becomes a vice if it is accompanied by an overly strong insistence that the simplest explanation is the only explanation. The data may have to be understood within some larger gestalt.

Sacrament. The outward and visible sign of an inward and spiritual grace.

Somatic. (Greek *soma*, body) A characteristic is said to be of *somatic* origin when the speaker wishes to emphasize that the characteristic was achieved by bodily change brought about during the lifetime of the individual by environmental impact or by practice.

Stochastic. (Greek, *stochazein*, to shoot with a bow at a target; that is, to scatter events in a partially random manner, some of which achieve a preferred outcome) If a sequence of events combines a random component with a selective process so that only certain outcomes of the random are allowed to endure, that sequence is said to be *stochastic.*

Stress. Lack of entropy, a condition arising when the external environment or internal sickness makes excessive or contradictory demands on an organism's ability to adjust. The organism lacks and needs *flexibility,* having used up its available uncommitted alternatives.

Tautology. An aggregate of linked propositions in which the validity of the *links* between them cannot be doubted. The truth of the propositions is not claimed. Example: Euclidean geometry.

Taxon. A unit or aggregate in the classification of animals or plants (e.g., a species, genus, or family).

Topology. A branch of mathematics that ignores quantities and deals only with the formal relations between components, especially components that can be represented geometrically. Topology deals with those characteristics (e.g., of a surface or body) that will remain unchanged under quantitative distortion.

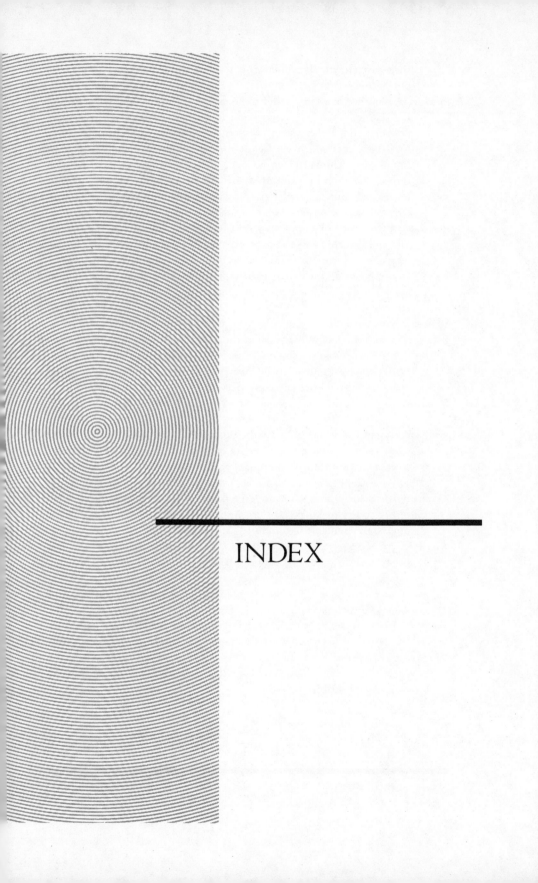

INDEX

abduction, 139, 142–144
Abraxas, 18
acclimation, 155–156
acquired characteristics, 149, 220
adaptation, 172–174, 184
addiction, 54, 139, 148, 172–174, 178
Aesop, 142
aesthetics, 8–9, 80, 211
Alice and flamingo, 178*n*
alternation of generations, 78
Alytes obstetricians (midwife toad), 151
Ames, A., Jr., 32–37
analogic change, 181
analogic coding, 111
Ancient Mariner, 210

Apollonian culture, 191
armaments races, 174
Ashby, R., 174
asymmetry, double, 10*n*
asymmetry in egg, 164
attributes, 60–61
Augustine, Saint, 2, 17, 52, 206
Aurignacian art, 141
autonomy, 126

Bacon, F., 217
Baer, K. E. von, 167–168
Balanoglossus, 181
Bateson, M. C., 11–12

parts and wholes, 9, 38–40, 93
Pasteur, L., 45
pattern, 8
 hierarchies of, 11
 and repetition, 29
pepper moth, 150
physical metaphors, 217
Plato, 4*n*, 170, 182
play, 125, 136, 138–139
pleroma, 7, 94
Plotinus, 2, 14
Pluto, 70–71
polyploidy, 55–56
population, 118*n*, 160, 222
power, myth of, 223
practice, 138, 195–196
prediction, 43, 44
presuppositions, 25, 143
Principia Mathematica, 116
probability, 44
processes, 184
proof, 27
Prospero, 14
Pryor, K., 122*n*
psychedelic experience, 43*n*
psychoanalysis, 14
purpose, 106, 207
Pythagoras, 51

quantity, 49–53

rain dance, 209
ratio, 53
recapitulation, 167, 180
recursiveness, 201
reductionism, 214
redwood forest, 112
relevance, 13
religious freedom, 218
repetition of parts, 9–10
rhythms, 80

rifle, 195
rigor, 183, 212, 219, 221
ritual, 137
rose, 50
Rosenblueth, A. N., 106
Rosetta stone, 46
runaway, 105
Russell, B., 19, 53*n*, 116, 123, 185, 196,
 199

sacrament, 6, 31
sacred, 213
schismogenesis, 105, 192
schizophrenia, 8, 139
schizothyme, 192
segmentation, 12
selection:
 external, 178
 internal, 177–178
self, 131, 139, 200
self-correction, 195
self-healing, 206
self-knowledge, 135–140
sexes, 77–79
Shaw, G. B., 188
Shiva, 17–18, 172, 208
shotgun, 195–196
situs inversus, 163
size, 9, 54–55
sociobiology, 133*n*
somatic change, 153–155
Spencer, H., 146, 180
Spencer-Brown, G., 91
spiral, 11–12, 164
stability, 61–62, 103
statistics, 44
steam engine, 110
 with governor, 43*n*, 105
Steno bredanensis, 121
Stevens, W., 77
stimulus, 100
stochastic process, 147